Introduction to
Process Safety for Undergraduates
and Engineers

This book is on in a series of process safety guidelines and concept books published by the Center for Chemical Process Safety (CCPS). Please go to www.wiley.com/go/ccps for a full list of titles in this series.

Introduction to Process Safety for Undergraduates and Engineers

CENTER FOR CHEMICAL PROCESS SAFETY
of the
AMERICAN INSTITUTE OF CHEMICAL ENGINEERS
New York, NY

Copyright © 2016 by the American Institute of Chemical Engineers, Inc. All rights reserved.

Published by John Wiley & Sons, Inc., Hoboken, New Jersey.

Published simultaneously in Canada.

No part of this publication may be reproduced, stored in a retrieval system, or transmitted in any form or by any means, electronic, mechanical, photocopying, recording, scanning, or otherwise, except as permitted under Section 107 or 108 of the 1976 United States Copyright Act, without either the prior written permission of the Publisher, or authorization through payment of the appropriate per-copy fee to the Copyright Clearance Center, Inc., 222 Rosewood Drive, Danvers, MA 01923, (978) 750-8400, fax (978) 750-4470, or on the web at www.copyright.com. Requests to the Publisher for permission should be addressed to the Permissions Department, John Wiley & Sons, Inc., 111 River Street, Hoboken, NJ 07030, (201) 748-6011, fax (201) 748-6008, or online at http://www.wiley.com/go/permission.

Limit of Liability/Disclaimer of Warranty: While the publisher and author have used their best efforts in preparing this book, they make no representations or warranties with respect to the accuracy or completeness of the contents of this book and specifically disclaim any implied warranties of merchantability or fitness for a particular purpose. No warranty may be created or extended by sales representatives or written sales materials. The advice and strategies contained herein may not be suitable for your situation. You should consult with a professional where appropriate. Neither the publisher nor author shall be liable for any loss of profit or any other commercial damages, including but not limited to special, incidental, consequential, or other damages.

For general information on our other products and services or for technical support, please contact our Customer Care Department within the United States at (800) 762-2974, outside the United States at (317) 572-3993 or fax (317) 572-4002.

Wiley also publishes its books in a variety of electronic formats. Some content that appears in print may not be available in electronic formats. For more information about Wiley products, visit our web site at www.wiley.com.

Library of Congress Cataloging-in-Publication Data is available.

ISBN: 978-1-118-94950-4

Printed in the United States of America.

10 9 8 7 6 5 4 3 2 1

It is our sincere intention that the information presented in this document will lead to an even more impressive safety record for the entire industry; however, neither the American Institute of Chemical Engineers (AIChE), its consultants, CCPS Technical Steering Committee and Subcommittee members, their employers, their employers officers and directors, warrant or represent, expressly or by implication, the correctness or accuracy of the content of the information presented in this document. As between (1) AIChE, its consultants, CCPS Technical Steering Committee and Subcommittee members, their employers, their employers officers and directors, and (2) the user of this document, the user accepts any legal liability or responsibility whatsoever for the consequence of its use or misuse.

CONTENTS

LIST OF TABLES xv
LIST OF FIGURES xvii
ACRONYMS AND ABBREVIATIONS xxi
GLOSSARY xxv
ACKNOWLEDGMENTS xxxiii
PREFACE xxxv
1. Introduction 1
   1.1 Purpose of this Handbook 1
   1.2 Target Audience 1
   1.3 Process Safety – What Is It? 1
   1.4 Organization of the Book 3
   1.5 References 4
2. Process Safety Basics 5
   2.1 Risk Based Process Safety 5
   Pillar: Commit to Process Safety 12
   2.2 Process Safety Culture 12
   2.3 Compliance with Standards 15
   2.4 Process Safety Competency 17
   2.5 Workforce Involvement 18
   2.6 Stakeholder Outreach 19
   Pillar: Understand Hazards and Risks 20
   2.7 Process Knowledge Management 20
   2.8 Hazard Identification and Risk Analysis 22
   Pillar: Manage Risk 25
   2.9 Operating Procedures 25
   2.10 Safe Work Practices 26
   2.11 Asset Integrity and Reliability 28

2.12 Contractor Management 30
2.13 Training And Performance Assurance 32
2.14 Management of Change 33
2.15 Operational Readiness 35
2.16 Conduct of Operations 37
2.17 Emergency Management 38
Pillar: Learn from Experience 42
2.18 Incident Investigation 42
2.19 Measurement and Metrics 45
2.20 Auditing 46
2.21 Management Review and Continuous Improvement 48
2.22 Summary 49
2.23 References 50

3. The Need for Process Safety 53
3.1 Process Safety Culture: BP Refinery Explosion, Texas City, 2005 58
    3.1.1 Summary 58
    3.1.2 Detailed Description 58
    3.1.3 Causes 59
    3.1.4 Key Lessons 61
    3.1.5 References and Links to Investigation Reports 63
3.2 Asset Integrity and Reliability: ARCO Channelview, Texas Explosion, 1990 64
    3.2.1 Summary 64
    3.2.2 Detailed Description 64
    3.2.3 Causes 65
    3.2.4 Key Lessons 65
    3.2.5 References and Links to Investigation Reports 65
3.3 Process Safety Culture: NASA Space Shuttle Columbia Disaster, 2003 66
    3.3.1 Summary 66
    3.3.2 Detailed Description 66

CONTENTS

- 3.3.3 Causes 68
- 3.3.4 Key Lessons 69
- 3.3.5 References and Links to Investigation Reports 70

3.4 Process Knowledge Management: Concept Sciences Explosion, Hanover Township PA, 1999 70
- 3.4.1 Summary 70
- 3.4.2 Detailed Description 70
- 3.4.3 Cause 72
- 3.4.4 Key Lessons 73
- 3.4.5 References and links to Investigation Reports 73

3.5 Hazard Identification and Risk Assessment: Esso Longford Gas Plant Explosion, 1998 73
- 3.5.1 Summary 73
- 3.5.2 Detailed Description 74
- 3.5.3 Cause 76
- 3.5.4 Key Lessons 76
- 3.5.5 References and Links to Investigation Reports 77

3.6 Operating Procedures: Port Neal, IA, Ammonium Nitrate Explosion, 1994 .. 77
- 3.6.1 Summary 77
- 3.6.2 Detailed Description 77
- 3.6.3 Causes 79
- 3.6.4 Key Lessons 80
- 3.6.5 References and Links to Investigation Reports 80

3.7 Safe Work Practices: Piper Alpha, North Sea, UK, 1988 80
- 3.7.1 Summary 80
- 3.7.2 Detailed Description 81
- 3.7.3 Causes 83
- 3.7.4 Key Lessons 84
- 3.7.5 References and Links to Investigation Reports 85

3.8 Contractor Management: Partridge Raleigh Oilfield Explosion, Raleigh, MS, 2006 85
    3.8.1 Summary 85
    3.8.2 Detailed Description 85
    3.8.3 Cause 86
    3.8.4 Key Lessons 86
    3.8.5 References and Links to Investigation Reports 88
3.9 Asset Integrity and Reliability: Explosion at Texaco Oil Refinery, Milford Haven, UK, 1994 88
    3.9.1 Summary 88
    3.9.2 Detailed Description 88
    3.9.3 Causes 89
    3.9.4 Key Lessons 90
    3.9.5 References and Links to Investigation Reports 91
3.10 Conduct of Operations: Formosa Plastics VCM Explosion, Illiopolis, IL, 2004 91
    3.10.1 Summary 91
    3.10.2 Detailed Description 91
    3.10.3 Causes 94
    3.10.4 Key Lessons 94
    3.10.5 References and Links to Investigation Reports 95
3.11 Management of Change: Flixborough Explosion, UK, 1974 95
    3.11.1 Summary 95
    3.11.2 Detailed Description 95
    3.11.3 Cause 98
    3.11.4 Key Lessons 98
    3.11.5 References and Links to Investigation Reports 99
3.12 Emergency Management: Sandoz Warehouse Fire, Switzerland, 1986 99
    3.12.1 Summary 99
    3.12.2 Key Lessons 101

CONTENTS xi

  3.12.3 References and links to investigation reports 102
 3.13 Conduct of Operations: Exxon Valdez, Alaska, 1989 102
  3.13.1 Summary 102
  3.13.2 Detailed Description 102
  3.13.3 Causes 105
  3.13.4 Key Lessons 105
  3.13.5 References and Links to Investigation Reports 106
 3.14 Compliance with Standards: Mexico City, PEMEX LPG Terminal, 1984 106
  3.14.1 Summary 106
  3.14.2 Detailed Description 106
  3.14.3 Causes 109
  3.14.4 Key Lessons 109
  3.14.5 References and Links to Investigation Reports 109
 3.15 Process Safety Culture: Methyl Isocyanate Release, Bhopal, India, 1984 110
  3.15.1 Summary 110
  3.15.2 Detailed Description 110
  3.15.3 Key Lessons 111
  3.15.4 References and Links to Investigation Reports 112
 3.16 Failure to Learn, BP Macondo Well Blowout, Gulf of Mexico, 2010 113
  3.16.1 Summary 113
  3.16.2 Detailed Description 113
  3.16.3 Key Lessons 118
  3.16.4 References and Links to Investigation Reports 119
 3.17 Summary 119
 3.18 References 120
4. Process Safety for Engineering Disciplines 121
 4.1 Introduction 121
 4.2 Process Knowledge Management 121

4.3 Compliance with Standards 124
4.4 Hazard Identification and Risk Analysis, Management Of Change 126
  Management of Organizational Change 127
4.5 Asset Integrity and Reliability 128
4.6 Safe Work Practices 129
4.7 Incident Investigation 130
4.8 Resources for Further Learning 130
4.8 Summary 132
4.9 References 132

5. Process Safety in Design 133
5.1 Process Safety Design Strategies 133
5.2 General Unit Operations and Their Failure Modes 134
  5.2.1 Pumps, Compressors, Fans 134
  5.2.2 Heat Exchange Equipment 141
  5.2.3 Mass Transfer; Distillation, Leaching and Extraction, Absorption 146
  5.2.4 Mechanical Separation / Solid-Fluid Separation 152
  5.2.5 Reactors and Reactive Hazards 158
  5.2.6 Fired Equipment 163
  5.2.7 Storage 167
5.3 Petroleum Processing 179
  5.3.1 General Process Safety Hazards in a Refinery 180
  5.3.2 Crude Handling and Separation 182
  5.3.3 Light Hydrocarbon Handling and Separation 183
  5.3.4 Hydrotreating 184
  5.3.5 Catalytic Cracking 185
  5.3.6 Reforming 187
  5.3.7 Alkylation 188
  5.3.8 Coking 190
5.4 Transient Operating States 192
  5.4.1 Overview 192

CONTENTS

    5.4.2 Example Process Safety Incidents 192

    5.4.3 Design Considerations 194

  5.5 References.. 194

6. Course Material 199

  6.1 Introduction.. 199

  6.2 Inherently Safer Design 199

  6.3 Process Safety Management and Conservation of Life 199

  6.4 Process Safety Overview and Safety in the Chemical Process Industries.. 200

  6.5 Process Hazards 201

    6.5.1 Chemical Reactivity Hazards.. 201

    6.5.2 Fires and Explosions 202

    6.5.3 Other Hazards 203

  6.6 Hazard Identification and Risk Analysis... 203

  6.7 Emergency Relief Systems 205

  6.8 Case Histories 206

    6.8.1 Runaway Reactions... 206

    6.8.2 Other Case Histories 207

  6.9 Other Modules 209

  6.10 Summary 209

  6.11 References... 209

7. Process Safety in the Workplace 211

  7.1 What to Expect.. 211

    7.1.1 Formal Training 211

    7.1.2 Interface with Operators, Craftsmen..................................... 214

  7.2 New Skills.. 215

    7.2.1 Non-Technical 215

    7.2.2 Technical... 216

  7.3 Safety Culture 217

  7.4 Conduct of Operations 218

    7.4.1 Operational Discipline 218

7.4.2 Engineering Discipline 230
7.4.3 Management Discipline 232
7.4.4 Other Conduct of Operations Topics for the New Engineer 237
7.5 Summary 238
7.6 References 238

APPENDIX A – EXAMPLE RAGAGEP LIST 241
APPENDIX B – LIST OF CSB VIDEOS 245
APPENDIX C – REACTIVE CHEMICALS CHECKLIST 249
C.1 Chemical Reaction Hazard Identification 249
C.2 Reaction Process Design Considerations 252
C.3 Resources and Publications 254

APPENDIX D – LIST OF SACHE COURSES 257
APPENDIX E – Reactivity Hazard Evaluation Tools 259
E.1 Screening Table and Flowchart 259
E.2 Reference 262

INDEX 263

LIST OF TABLES

Table 2.1. Comparison of RBPS elements to OSHA PSM elements. 10
Table 2.2. Examples and sources of process safety related standards, codes, regulations, and laws. 16
Table 2.3 Hazard evaluation synonyms 23
Table 2.4 Typical HAZOP review table format. 24
Table 2.5. Activities typically included in the scope of the safe work element 28
Table 3.1 Selected incidents and Process Safety Management systems. 54
Table 4.1. Process safety activities for new engineers. 122
Table 4.2. Incidents with organizational change involvement 128
Table 5.1 Common failure modes, causes, consequences, design considerations for fluid transfer equipment 140
Table 5.2 Common failure modes, causes, consequences, design considerations for heat exchange equipment. 147
Table 5.2 Common failure modes, causes, consequences, design considerations for heat exchange equipment, continued. 148
Table 5.3 Common failure modes, causes, consequences, design considerations for reactors 164
Table 7.1 Example simplified process safety training class matrix. 212
Table A-1. RAGAGEP List for XYZ Chemicals 241
Table B.1 List of CSB Videos 245
Table D.1 List of SACHE Courses 257
Table E.1 Example Form to Document Screening of Chemical Reactivity Hazards 259

LIST OF FIGURES

Figure 2.1. Picture of a nitroglycerine reactor in the 19th century. 7
"Alfred Nobel in Scotland". Nobelprize.org. Nobel Media AB 2014. Web. 15 Sep 2015. <http://www.nobelprize.org/alfred_nobel/biographical/articles/dolan/>....... 7
Figure 2.2. Continuous nitroglycerine reactor, courtesy Biazzi SA (www.Biazzi.com). 7
Figure 2.3. Illustration of risk. 11
Figure 2.4. Challenger Disaster, courtesy NASA. 12
Figure 2.5. Building damage and charge tank crater, Hydroxylamine explosion, courtesy CSB. 20
Figure 2.6. Collapsed tank at Motiva refinery, courtesy CSB. 27
Figure 2.7. Rupture in 52-inch component of line, courtesy CSB. 29
Figure 2.8. Aerial view of the burning Monsanto plant after the 1947 Texas City Disaster, (http://texashistory.unt.edu/ark:/67531/metapth11883) University of North Texas Libraries, The Portal to Texas History, crediting Moore Memorial Public Library, Texas City, Texas. 39
Figure 2.9. CCPS and API Process Safety Metric Pyramid (Ref. 2.46). 44
Figure 2.10. Photograph of failed end of heat exchanger, (Ref. 2.33). 47
.... 57
Figure 3.1. Swiss Cheese model of incidents, Ref. 3.1. 57
Figure 3.2. Process flow diagram of the Raffinate Column and blowdown drum, source (CCPS, 2008). 59
Figure 3.3. Texas City Isom Unit aftermath, courtesy CSB. 60
Figure 3.4. Portable buildings destroyed where contractors were located, courtesy CSB. 60
Figure 3.5. Process flow diagram of wastewater tank. 64
Figure 3.6. Columbia breaking up, courtesy NASA. 67
Figure 3.7. A shower of foam debris after the impact on Columbia's left wing. The event was not observed in real time, courtesy NASA. 67
Figure 3.8. Damage to Concept Sciences Hanover Facility, courtesy Tom Volk, The Morning Call. 71
Figure 3.9. Simplified process flow diagram of the CSI HA vacuum distillation process, courtesy CSB. 72
Figure 3.10. Simplified schematic of absorber, (CCPS, 2008). 74
Figure 3.11. Simplified schematic of the gas plant (CCPS, 2008) 75
Figure 3.12 Neutralizer and rundown tank, source, (EPA, 1996). 78
Figure 3.13. AN plant area after explosion, source, (EPA 1996). 79

Figure 3.14. Piper Alpha platform, source (CCPS, 2008). 81
Figure 3.15. Schematic of Piper Alpha platform, source (CCPS, 2008) 82
Figure 3.16. Tanks involved in the Partridge Raleigh oilfield explosion, source (CSB, 2006). 86
Figure 3.17. Tank 3 lid, source (CSB, 2007). 87
Figure 3.18. Ref. (CCPS, 2008) Picture courtesy of Western Mail and Echo Ltd.89
Figure 3.19. The 30 inch flare line elbow that failed and released 20 tons of vapor, source (HSE, 1994). 90
Figure 3.20. Smoke plumes from Formosa plant, source (CSB 2007). 92
Figure 3.21. Reactor building elevation view, source (CSB 2007) 92
Figure 3.22. Cutaway of the reactor building, source (CSB 2007). 93
Figure 3.23. Schematic of Flixborough piping replacement, source Report of the Court of Inquiry. 96
Figure 3.24. The collapsed 20 inch pipe. 97
Figure 3.25. Damage to Flixborough plant. 98
Figure 3.26. Damage to Flixborough control room. 98
Figure 3.27. Sandoz Warehouse firefighting efforts, source (CCPS, 2008) 100
Figure 3.28. Impact of Sandoz Warehouse firewater runoff, (CCPS, 2008). 101
Figure 3.29. Exxon Valdez tanker leaking oil, courtesy of Exxon Valdez Oil Spill Trustee Council. 103
Figure 3.30. Oiled loon onshore, courtesy of Exxon Valdez Oil Spill Trustee Council. 103
Figure 3.31. Aerial of a maxi-barge with water tanks and spill works hosing a beach, Prince William Sound, courtesy of Exxon Valdez Oil Spill Trustee Council. 104
Figure 3.32. Cleanup workers spray oiled rocks with high pressure hoses, courtesy of Exxon Valdez Oil Spill Trustee Council. 104
Figure 3.33 Layout of PEMEX LPG Terminal, source, CCPS, 2008) 107
Figure 3.34. PEMEX LPG Terminal prior to explosion source, CCPS, 2008. 108
Figure 3.35. PEMEX LPG Terminal after the explosion source, CCPS, 2008. ... 108
Figure 3.36. Schematic of emergency relief effluent treatment system that included a scrubber and flare tower in series, source AIChE. 111
Figure 3.37. Photograph taken shortly after the incident. A pipe rack is shown on the left and the partially buried storage tanks (three total) for MIC are located in the center of the photo right, (source Willey 2006). 112
Figure 3.38. Fire on Deepwater Horizon, source (CSB, 2010). 114
Figure 3.39. Location of Mud-Gas separator, source (TO, 2011) 115
Figure 3.40. Gas release points, source (TO, 2011). 116
Figure 3.41. Macondo Well blowout preventer, source (CSB 2010) 117
Figure 5.1. Damage from fire caused by mechanical seal failure. 135

LIST OF FIGURES xix

Figure 5.2. Pump explosion from running isolated. 136
Figure 5.3. Schematic of centrifugal pump, Ref. 5.6. 137
Figure 5.4. Single and Double Mechanical Seals, Ref. 5.7. 137
Figure 5.5. Two-screw type PD Pump, courtesy Colfax Fluid Handling. 139
Figure 5.6. Rotary Gear PD pump, source http://www.tpub.com/gunners/99.htm. 139
Figure 5.7. Example application data sheet, courtesy of OEC Fluid Handling.... 142
Figure 5.8. Ruptured pipe from reaction with heat transfer fluid. 143
Figure 5.9. Shell and tube heat exchanger, Ref. 5.9. 144
Figure 5.10. Cutaway drawing of a Plate-and-Frame Heat Exchanger, Ref. 5.10 145
Figure 5.10. Schematic of air cooled heat exchanger, Ref. 5.11 145
Figure 5.12. Double tube sheet, courtesy www.wermac.org 146
Figure 5.13. A. Example distillation column schematic Ref. 5.11, and B. typical industrial distillation column, ©Sulzer Chemtech Ltd. 148
Figure 5.14. Schematic of carbon bed adsorber system, Ref. 5.16. 150
Figure 5.15. Damage to dust collector bags, Ref. 5.25 154
Figure 5.16. Tube sheet of dust collector, Ref. 5.25. 155
Figure 5.17. A horizontal peeler centrifuge with a Clean-In-Place system and a discharge chute, (Ref. 5.26). 156
Figure 5.18. Cross sectional view of a continuous pusher centrifuge (Ref 5.26). 156
Figure 5.19. Schematic of baghouse, courtesy Donaldson-Torit. 157
Figure 5.20. Dust collector explosion venting, courtesy Fike 157
Figure 5.21. Seveso Reactor, adapted from SACHE presentation by Ron Willey. 159
Figure 5.22. T2 Laboratories site before and after the explosion, Ref. 5.28. 160
Figure 5.23. T2 Laboratories blast, Ref. 5.28. 161
Figure 5.24. Portion of 3 inch thick reactor, Ref. 5.28. 161
Figure 5.25. Damaged heater, Example 1. 165
Figure 5.26. Heater and adjacent column at NOVA Bayport plant, Example 2. . 165
Figure 5.27. Buncefield before the explosion and fires, Ref. 5.32. 169
Figure 5.28. Buncefield after the explosion and fires, Ref 5.32. 169
Figure 5.29. Molasses tank failure; before and after. 170
Figure 5.30. 1) Pipe connections in panel 2) Chemfos 700 and Liq. Add lines... 170
Figure 5.31. Cloud of nitric oxide and nitrogen dioxide. 171
Figure 5.32. Tank collapsed by vacuum. 172
Figure 5.33. Schematic diagram of UST leak detection methods, courtesy EPA, Ref. 5.36. 172
Figure 5.34. Mounded underground tank, courtesy BNH Gas Tanks. 173

Figure 5.35. Schematics of external (a) and internal floating (b) roof tanks, courtesy of petroplaza.com. 174
Figure 5.36 Pressurized gas storage tank. 176
Figure 5.37. Refinery flow diagram, Ref. 43. 180
Figure 5.39. Atmospheric separation process flow diagram, courtesy OSHA...... 182
Figure 5.40. Hydrotreater process flow diagram, Ref. 5.43. 184
Figure 5.41. Fluid Catalytic Cracking (FCC) process flow diagram, Ref. 41....... 186
Figure 5.42. CCR Naphtha Reformer process flow diagram, Ref. 43. 187
Figure 5.46. HF Alkylation process flow diagram. Ref. 5.46. 189
Figure 5.44. Process flow diagram for a delayed coker unit, Ref. 5.43. 190
Figure 5.45. Polymer catch tank, Ref. 5.50... 193
Figure 7.1. Car Seal on a valve handle. Seal can be broken in an emergency if necessary to change the position of a valve, courtesy 228

ACRONYMS AND ABBREVIATIONS

ACC	American Chemistry Council
AIChE	American Institute of Chemical Engineers
API	American Petroleum Institute
ASME	American Society of Mechanical Engineers
BLEVE	Boiling Liquid Expanding Vapor Explosion
BMS	Burner Management System
CEI	Chemical Exposure Index (Dow Chemical)
CFR	Code of Federal Registry
CMA	Chemical Manufacturers Association
CSB	US Chemical Safety and Hazard Investigation Board
CCPS	Center for Chemical Process Safety
CCR	Continuous Catalyst Regeneration
COO	Conduct of Operations
CPI	Chemical Process Industries
DCU	Delayed Coker Unit
DDT	Deflagration to Detonation Transition
DIERS	Design Institute for Emergency Relief Systems
ERS	Emergency Relief System
EPA	US Environmental Protection Agency
FCCU	Fluidized Catalytic Cracking Unit
F&EI	Fire and Explosion Index (Dow Chemical)
FMEA	Failure Modes and Effect Analysis
HAZMAT	Hazardous Materials
HAZOP	Hazard and Operability Study

HIRA	Hazard Identification and Risk Analysis	
HTHA	High Temperature Hydrogen Attack	
HSE	Health & Safety Executive (UK)	
I&E	Instrument and Electrical	
IDLH	Immediately Dangerous to Life and Health	
ISD	Inherently Safer Design	
ISO	International Organization for Standardization	
ISOM	Isomerization Unit	
ITPM	Inspection Testing and Preventive Maintenance	
LFL	Lower Flammable Limit	
LNG	Liquefied Natural Gas	
LOPA	Layer of Protection Analysis	
LOTO	Lock Out Tag Out	
LPG	Liquefied Petroleum Gas	
MAWP	Maximum Allowable Working Pressure	
MCC	Motor Control Center	
MIE	Minimum Ignition Energy	
MOC	Management of Change	
MOOC	Management of Organizational Change	
MSDS	Material Safety Data Sheet	
NASA	National Aeronautics and Space Administration	
NDT	Non Destructive Testing	
NFPA	National Fire Protection Association	
OCM	Organizational Change Management	
OIMS	Operational Integrity Management System (ExxonMobil)	
OSHA	US Occupational Safety and Health Administration	
PHA	Process Hazard Analysis	
PLC	Programmable Logic Controller	

ACRONYMS AND ABBREVIATIONS

PRA	Probabilistic Risk Assessment
PRD	Pressure Relief Device
PRV	Pressure Relief Valve
PSB	Process Safety Beacon
PSE	Process Safety Event
PSI	Process Safety Information
PSI	Process Safety Incident
PSM	Process Safety Management
PSO	Process Safety Officer
PSSR	Pre-Startup Safety Review
QRA	Quantitative Risk Analysis
RBPS	Risk Based Process Safety
RAGAGEP	Recognized and Generally Accepted Good Engineering Practice
RMP	Risk Management Plan
SACHE	Safety and Chemical Engineering Education
SCAI	Safety Controls Alarms and Interlocks
SHE	Safety, Health and Environmental (sometimes written as EHS or HSE)
SHIB	Safety Hazard Information Bulletin
SIS	Safety Instrumented Systems
SME	Subject Matter Expert
TQ	Threshold Quantity
UFL	Upper Flammable Limit
UK	United Kingdom
US	United States
UST	Underground Storage Tank

GLOSSARY

Asset integrity A PSM program element involving work activities that help ensure that equipment is properly designed, installed in accordance with specifications, and remains fit for purpose over its life cycle. Also called asset integrity and reliability.

Atmospheric Storage Tank A storage tank designed to operate at any pressure between ambient pressure and 0.5 psig (3.45kPa gage).

Boiling-Liquid-Expanding-Vapor Explosion (BLEVE) A type of rapid phase transition in which a liquid contained above its atmospheric boiling point is rapidly depressurized, causing a nearly instantaneous transition from liquid to vapor with a corresponding energy release. A BLEVE of flammable material is often accompanied by a large aerosol fireball, since an external fire impinging on the vapor space of a pressure vessel is a common cause. However, it is not necessary for the liquid to be flammable to have a BLEVE occur.

Checklist Analysis A hazard evaluation procedure using one or more pre-prepared lists of process safety considerations to prompt team discussions of whether the existing safeguards are adequate.

Chemical Process Industry The phrase is used loosely to include facilities which manufacture, handle and use chemicals.

Combustible Dust Any finely divided solid material that is 420 microns or smaller in diameter (material passing through a U.S. No. 40 standard sieve) and presents a fire or explosion hazard when dispersed and ignited in air or other gaseous oxidizer.

Conduct of Operations (COO) The embodiment of an organization's values and principles in management systems that are developed, implemented, and maintained to (1) structure operational tasks in a manner consistent with the organization's risk tolerance, (2) ensure that every task is performed deliberately and correctly, and (3) minimize variations in performance.

Explosion A release of energy that causes a pressure discontinuity or blast wave.

Failure Mode and Effects Analysis	A hazard identification technique in which all known failure modes of components or features of a system are considered in turn, and undesired outcomes are noted.
Flammable Liquids	Any liquid that has a closed-cup flash point below 100 °F (37.8 °C), as determined by the test procedures described in NFPA 30 and a Reid vapor pressure not exceeding 40 psia (2068.6 mm Hg) at 100°F (37.8 °C), as determined by ASTM D 323, Standard Method of Test for Vapor Pressure of Petroleum Products (Reid Method). Class IA liquids shall include those liquids that have flash points below 73 °F (22.8 °C) and boiling points below 100 F (37.8 °C). Class IB liquids shall include those liquids that have flash points below 73°F (22.8 °C) and boiling points at or above 100 °F (37.8 °C). Class IC liquids shall include those liquids that have flash points at or above 73 °F (22.8 °C), but below 100 °F (37.8 °C). (NFPA 30).
Hazard Analysis	The identification of undesired events that lead to the materialization of a hazard, the analysis of the mechanisms by which these undesired events could occur and usually the estimation of the consequences.
Hazard and Operability Study (HAZOP)	A systematic qualitative technique to identify process hazards and potential operating problems using a series of guide words to study process deviations. A HAZOP is used to question every part of a process to discover what deviations from the intention of the design can occur and what their causes and consequences may be. This is done systematically by applying suitable guide words. This is a systematic detailed review technique, for both batch and continuous plants, which can be applied to new or existing processes to identify hazards
Hazard Identification	The inventorying of material, system, process and plant characteristics that can produce undesirable consequences through the occurrence of an incident.
Hazard Identification and Risk Analysis (HIRA)	A collective term that encompasses all activities involved in identifying hazards and evaluating risk at facilities, throughout their life cycle, to make certain that risks to employees, the public, or the environment are consistently controlled within the organization's risk tolerance.
Hot Work	Any operation that uses flames or can produce sparks (e.g., welding).

GLOSSARY

Incident
An event, or series of events, resulting in one or more undesirable consequences, such as harm to people, damage to the environment, or asset/business losses. Such events include fires, explosions, releases of toxic or otherwise harmful substances, and so forth.

Incident Investigation
A systematic approach for determining the causes of an incident and developing recommendations that address the causes to help prevent or mitigate future incidents. See also Root cause analysis and Apparent cause analysis.

Interlock
A protective response which is initiated by an out-of-limit process condition. Instrument which will not allow one part of a process to function unless another part is functioning. A device such as a switch that prevents a piece of equipment from operating when a hazard exists. To join two parts together in such a way that they remain rigidly attached to each other solely by physical interference. A device to prove the physical state of a required condition and to furnish that proof to the primary safety control circuit.

Layer of Protection Analysis (LOPA)
An approach that analyzes one incident scenario (cause-consequence pair) at a time, using predefined values for the initiating event frequency, independent protection layer failure probabilities, and consequence severity, in order to compare a scenario risk estimate to risk criteria for determining where additional risk reduction or more detailed analysis is needed. Scenarios are identified elsewhere, typically using a scenario-based hazard evaluation procedure such as a HAZOP Study.

Lockout/Tagout
A safe work practice in which energy sources are positively blocked away from a segment of a process with a locking mechanism and visibly tagged as such to help ensure worker safety during maintenance and some operations tasks.

Management of Change (MOC)
A system to identify, review and approve all modifications to equipment, procedures, raw materials and processing conditions, other than "replacement in kind," prior to implementation.

Management System
A formally established set of activities designed to produce specific results in a consistent manner on a sustainable basis.

Mechanical Integrity
A management system focused on ensuring that equipment is designed, installed, and maintained to perform the desired function.

Near-Miss	An unplanned sequence of events that could have caused harm or loss if conditions were different or were allowed to progress, but actually did not.
Operating Procedures	Written, step-by-step instructions and information necessary to operate equipment, compiled in one document including operating instructions, process descriptions, operating limits, chemical hazards, and safety equipment requirements.
Operational Discipline (OD)	The performance of all tasks correctly every time; Good OD results in performing the task the right way every time. Individuals demonstrate their commitment to process safety through OD. OD refers to the day-to-day activities carried out by all personnel. OD is the execution of the COO system by individuals within the organization.
Operational Readiness	A PSM program element associated with efforts to ensure that a process is ready for start-up/restart. This element applies to a variety of restart situations, ranging from restart after a brief maintenance outage to restart of a process that has been mothballed for several years.
Organizational Change	Any change in position or responsibility within an organization or any change to an organizational policy or procedure that affects process safety.
Organizational Change Management (OCM)	A method of examining proposed changes in the structure or organization of a company (or unit thereof) to determine whether they may pose a threat to employee or contractor health and safety, the environment, or the surrounding populace.
OSHA Process Safety Management (OSHA PSM)	A U.S. regulatory standard that requires use of a 14-element management system to help prevent or mitigate the effects of catastrophic releases of chemicals or energy from processes covered by the regulations 49 CFR 1910.119.
Pressure Relief Valve (PRV)	A pressure relief device which is designed to reclose and prevent the further flow of fluid after normal conditions have been restored.
Pressure Safety Valve (PSV)	See Pressure Relief Valve

GLOSSARY

Pre-Startup Safety Review (PSSR)
A systematic and thorough check of a process prior to the introduction of a highly hazardous chemical to a process. The PSSR must confirm the following: Construction and equipment are in accordance with design specifications; Safety, operating, maintenance, and emergency procedures are in place and are adequate; A process hazard analysis has been performed for new facilities and recommendations and have been resolved or implemented before startup, and modified facilities meet the management of change requirements; and training of each employee involved in operating a process has been completed.

Preventive Maintenance
Maintenance that seeks to reduce the frequency and severity of unplanned shutdowns by establishing a fixed schedule of routine inspection and repairs.

Probabilistic Risk Assessment (PRA)
A commonly used term in the nuclear industry to describe the quantitative evaluation of risk using probability theory.

Process Hazard Analysis (PHA)
An organized effort to identify and evaluate hazards associated with processes and operations to enable their control. This review normally involves the use of qualitative techniques to identify and assess the significance of hazards. Conclusions and appropriate recommendations are developed. Occasionally, quantitative methods are used to help prioritize risk reduction.

Process Knowledge Management
A Process Safety Management (PSM) program element that includes work activities to gather, organize, maintain, and provide information to other PSM program elements. Process safety knowledge primarily consists of written documents such as hazard information, process technology information, and equipment-specific information. Process safety knowledge is the product of this PSM element.

Process Safety Culture
The common set of values, behaviors, and norms at all levels in a facility or in the wider organization that affect process safety.

Process Safety Incident/Event
An event that is potentially catastrophic, i.e., an event involving the release/loss of containment of hazardous materials that can result in large-scale health and environmental consequences.

Process Safety Information (PSI)
Physical, chemical, and toxicological information related to the chemicals, process, and equipment. It is used to document the configuration of a process, its characteristics, its limitations, and as data for process hazard analyses.

Process Safety Management (PSM)	A management system that is focused on prevention of, preparedness for, mitigation of, response to, and restoration from catastrophic releases of chemicals or energy from a process associated with a facility.
Process Safety Management Systems	Comprehensive sets of policies, procedures, and practices designed to ensure that barriers to episodic incidents are in place, in use, and effective.
Reactive Chemical	A substance that can pose a chemical reactivity hazard by readily oxidizing in air without an ignition source (spontaneously combustible or peroxide forming), initiating or promoting combustion in other materials (oxidizer), reacting with water, or self-reacting (polymerizing, decomposing or rearranging). Initiation of the reaction can be spontaneous, by energy input such as thermal or mechanical energy, or by catalytic action increasing the reaction rate.
Recognized and Generally Accepted Good Engineering Practice (RAGAGEP)	A term originally used by OSHA, stems from the selection and application of appropriate engineering, operating, and maintenance knowledge when designing, operating and maintaining chemical facilities with the purpose of ensuring safety and preventing process safety incidents.
	It involves the application of engineering, operating or maintenance activities derived from engineering knowledge and industry experience based upon the evaluation and analyses of appropriate internal and external standards, applicable codes, technical reports, guidance, or recommended practices or documents of a similar nature. RAGAGEP can be derived from singular or multiple sources and will vary based upon individual facility processes, materials, service, and other engineering considerations.
Responsible Care©	An initiative implemented by the Chemical Manufacturers Association (CMA) in 1988 to assist in leading chemical processing industry companies in ethical ways that increasingly benefit society, the economy and the environment while adhering to ten key principles.
Risk Management Program (RMP) Rule	EPA's accidental release prevention Rule, which requires covered facilities to prepare, submit, and implement a risk management plan.

Risk-Based Process Safety (RBPS)	The Center for Chemical Process Safety's (CCPS) PSM system approach that uses risk-based strategies and implementation tactics that are commensurate with the risk-based need for process safety activities, availability of resources, and existing process safety culture to design, correct, and improve process safety management activities.
Safety Instrumented System (SIS)	The instrumentation, controls, and interlocks provided for safe operation of the process.
Vapor Cloud Explosion (VCE)	The explosion resulting from the ignition of a cloud of flammable vapor, gas, or mist in which flame speeds accelerate to sufficiently high velocities to produce significant overpressure.

ACKNOWLEDGMENTS

The American Institute of Chemical Engineers (AIChE) and the Center for Chemical Process Safety (CCPS) express their appreciation and gratitude to all members of the *Introduction to Process Safety for Undergraduates and Engineers* and their CCPS member companies for their generous support and technical contributions in the preparation of this book.

Subcommittee Members:

Don Abrahamson	CCPS - Emeritus
Iclal Atay	New Jersey DEP
Brooke Cailleteau	LyondellBasell (Houston Refining)
Dan Crowl	Michigan Technical University
Jerry Forest	Celanese - Project Chair
Robert Forest	University of Delaware
Jeff Fox	Dow Corning
Mikelle Moore	Buckman North America
Albert Ness	CCPS – Process Safety Writer
Eric Peterson	MMI Engineering
Robin Pitblado	DNV GL
Dan Sliva	CCPS - Staff Consultant
Rob Smith	Siemens Consulting
Scott Wallace	Olin Corporation

The collective industrial experience and know-how of the subcommittee members plus these individuals makes this book especially valuable to engineers who develop and manage process safety programs and management systems, including the identification of the competencies needed to create and maintain these systems.

The book committee wishes to express their appreciation to Albert Ness and of CCPS and Arthur Baulch of the AIChE for their contributions in preparing this book for publication.

Before publication, all CCPS books are subjected to a thorough peer review process. CCPS gratefully acknowledges the thoughtful comments and suggestions of the peer reviewers. Their work enhanced the accuracy and clarity of these guidelines.

Peer Reviewers:

John Alderman	Hazard and Risk Analysis, LLC
Dan Crowl	Professor of Chemical Engineering, Michigan Technical University, Retired
Dr. Kerry M. Dooley	BASF Professor of Chemical Engineering, Louisiana State University
John Herber	CCPS Staff Consultant
Greg Hounsell	CCPS Staff Consultant
Robert W. Johnson	President, Unwin Company
Jerry Jones	CCPS Staff Consultant
Michael L. LaFond	Engineer, Hemlock Semiconductor/Dow Corning
Robert J. Lovelett	Chemical Engineering Student, University of Delaware
John Murphy	CCPS Staff Consultant
Eloise Roche	Dow Chemical
Robert Rosen	RRS Engineering
Chad Schaeffer	Chemical Engineering Student, University of Delaware
Steve Selk	Department of Homeland Security
Chris Tagoe	VP HES, Cameron
Bruce Vaughen	Principal Consultant, BakerRisk
Ron Wiley	Professor of Chemical Engineering, Northeastern University
John Zondlo	Professor of Chemical Engineering, West Virginia University
Lucy Yi	CCPS – China Section

Although the peer reviewers have provided many constructive comments and suggestions, they were not asked to endorse this book and were not shown the final manuscript before its release.

PREFACE

The Center for Chemical Process Safety (CCPS) was created by the AIChE in 1985 after the chemical disasters in Mexico City, Mexico, and Bhopal, India. The CCPS is chartered to develop and disseminate technical information for use in the prevention of major chemical accidents. The Center is supported by more than 180 chemical process industries (CPI) sponsors who provide the necessary funding and professional guidance to its technical committees. The major product of CCPS activities has been a series of guidelines to assist those implementing various elements of a process safety and risk management system. This book is part of that series.

The AIChE has been closely involved with process safety and loss control issues in the chemical and allied industries for more than five decades. Through its strong ties with process designers, constructors, operators, safety professionals, and members of academia, AIChE has enhanced communications and fostered continuous improvement of the industry's high safety standards. AIChE publications and symposia have become information resources for those devoted to process safety and environmental protection.

The integration of process safety into the engineering curricula is an ongoing goal of the CCPS. To this end, CCPS created the Safety and Chemical Engineering Education (SACHE) committee which develops training modules for process safety. One textbook covering the technical aspects of process safety for students already exists; however, there is no textbook covering the concepts of process safety management and the need for process safety for students. The CCPS Technical Steering Committee initiated the creation of this book to assist colleges and universities in meeting this challenge and to aid Chemical Engineering programs in meeting recent accreditation requirements for including process safety into the chemical engineering curricula.

1

Introduction

1.1 Purpose of this Handbook

This book is intended to be used as a reference material for either a stand-alone process safety course or as supplemental material for existing curricula. This book is not a technical book; rather, the intent of the material is to familiarize the student or an engineer new to process safety with:

- The concept of process safety management (PSM).
- The 20 elements of process safety defined by the Center for Chemical Process Safety (CCPS).
- The need for process safety as illustrated by examples of major process safety incidents that have occurred.
- Process safety tasks for other engineering disciplines.
- Process safety concerns with some selected unit operations.
- Show how various aspects of process safety have a direct tie-in to existing chemical engineering curricula.
- Describe the many tasks that can be expected of an engineer new to process safety with respect to process safety in their first few years on the job.

1.2 Target Audience

This primary audience for this publication is junior to graduate level Chemical Engineering students and those entering the workforce and engineers new to process safety. However, since there are no technical pre-requisites recommended, it may also be used by other engineering disciplines at similar levels.

1.3 Process Safety – What Is It?

In the chemical, petrochemical and most other industries, you will find that all companies are required to have an occupational safety program, with a focus on personal safety (this program may be required by regulations in many countries, states and local areas. It can apply to workers in a manufacturing plant, a research laboratory or pilot plant, and even to office locations). That program is going to focus on personal safety. The focus of these programs is to prevent harm to workers from workplace accidents such as falls, cuts, sprains and strains, being struck by objects, repetitive motion injuries, and so on. They are good and in fact, very necessary programs. They are not, however, what *Process Safety* is about.

Process Safety is defined as "a discipline that focuses on the prevention of fires, explosions, and accidental chemical releases at chemical process facilities". Such events don't only happen at chemical facilities, they occur in refineries, off-shore drilling facilities, etc. Another definition is that process safety is about the prevention of, preparedness for, mitigation of, response to, or restoration from catastrophic releases of chemicals or energy from a process associated with a facility.

After an explosion in a BP Texas City refinery in 2005 that killed 15 people and injured over 170 others, an independent commission was created to examine the process safety mind-set, or culture, of BPs refinery operations, this commission came to be known as the Baker Panel. The Baker Panel said this about process safety:

> "Process safety hazards can give rise to major accidents involving the release of potentially dangerous materials, the release of energy (such as fires and explosions), or both. Process safety incidents can have catastrophic effects and can result in multiple injuries and fatalities, as well as substantial economic, property, and environmental damage. Process safety refinery incidents can affect workers inside the refinery and members of the public who reside nearby. Process safety in a refinery involves the prevention of leaks, spills, equipment malfunctions, over-pressures, excessive temperatures, corrosion, metal fatigue, and other similar conditions. Process safety programs focus on the design and engineering of facilities, hazard assessments, management of change, inspection, testing, and maintenance of equipment, effective alarms, effective process control, procedures, training of personnel, and human factors." (Ref 1.1)

The term "refinery" in that paragraph can be replaced by "petrochemical plant", "chemical process facility", "solids handling facility", "water treatment plants", "ammonia refrigeration plants", "off-shore operations" or any number of terms for a plant that handles or processes flammable, combustible, toxic, or reactive materials. For the rest of this book, the term process facility or just facility will be used to mean the previously mentioned facilities and any other operation that handles or processes flammable, combustible, toxic, or reactive materials.

The quote from the Baker report states that process safety is not limited to the operation of a facility. During the basic research and process research phases, process safety programs cover the operation of pilot facilities. They also cover the selection of the chemistry and unit operations chosen to achieve the design intent of the process. During the design and engineering phase, process safety is involved in choices about what type of unit operations and equipment items to use, the facility layout, and so on. Running a facility involves, as was mentioned above, "hazard assessments, management of change, inspection, testing, and maintenance of equipment, effective alarms, effective process control, procedures, and training of personnel". The choices made about process features during research and development and pilot work can make these activities easier or more difficult.

1.4 Organization of the Book

Chapter 2 gives a brief history of process safety and of process safety management. The evolution of process safety management principles from the initial twelve elements of process safety management developed by CCPS, and the process regulatory framework of the Occupational Safety and Health Administration's (OSHA) PSM regulations to the current twenty elements of the CCPS Risk Based Process Safety (RBPS) management system is discussed.

Chapter 3 describes several process safety incidents that demonstrate the need for a good PSM system. Each incident is described, and then the relevance of a few relevant RBPS elements are listed.

Chapter 4 describes the role of several engineering disciplines, Chemical, Mechanical, Civil, Instrumentation and Electrical (I&E) Engineers, and Safety Engineers with respect to how new engineers will be involved in process safety. PSM is a team effort between many disciplines.

Chapter 5 covers a few key process safety concerns with respect to some unit operations and equipment found in the chemical, biochemical and petrochemical and industries that could handle hazardous materials. Combinations of these unit operations are many and varied across the process industries. In the petrochemical industry there are several common operations that are used, and this book describes the process safety concerns of some of those operations. This chapter also introduces the concept of Inherent Safety (IS) and Inherently Safer Design (ISD). ISD focuses on eliminating or reducing hazards inherent in a process as opposed to trying to manage the hazards.

Chapter 6 lists training modules available from the Safety and Chemical Engineering Education (SACHE) Committee through the AIChE and describes the courses and their relevance to some Chemical Engineering courses. This chapter can be used as a guide for supplementing existing courses.

Chapter 7 describes process safety related duties that a new engineer can expect to encounter during the first year to two years in the process industry. For a PSM system to work well, all people involved in the process must execute their roles and responsibilities in a deliberate and structured manner to achieve a high level of human performance. This is called Conduct of Operations. Chapter 7 describes many tasks of engineers with respect to Conduct of Operations, as well as what the engineer should expect operators, maintenancde and management with respect to their roles.

1.5 References

1.1 The Report of the BP U.S. Refineries Independent Safety Review Panel, January 2007. (http://www.bp.com/liveassets/bp_internet/globalbp/globalbp_uk_english/SP/STAGING/local_assets/assets/pdfs/Baker_panel_report.pdf).

2

Process Safety Basics

2.1 Risk Based Process Safety

In Chapter 1 you were introduced to the concept of *process safety*. This chapter is going to cover a brief history of process safety plus the concepts of *management systems* and *risk based process safety,* along with a description of the elements of a *risk based process safety management system* as proposed by the Center for Chemical Process Safety (CCPS) in 2007.

    History of Process Safety. Organizations in the process industries have a long standing concern for process safety. (See the inset about the manufacture of nitroglycerine as an example.) Organizations originally had safety reviews for processes that relied on the experience and expertise of the people in the review. In the middle of the 20th century, more formal review techniques began to appear in the process industries. These included the Hazard and Operability (HAZOP) review, developed by ICI in the 1960s, Failure Mode and Effect Analysis (FMEA), Checklist and What-If reviews. These were qualitative techniques for assessing the hazards of a process.

    Quantitative analysis techniques, such as Fault Tree Analysis (FTA), which had been in use by the nuclear industry, Quantitative Risk Assessment (QRA), and Layer of Protection Analysis (LOPA) also began to be used in the process industries in the 1970s, 1980s and 1990s. Modeling techniques were developed for analyzing the consequences of spills and releases, explosions, and toxic exposures. The Design Institute of Emergency Relief Systems (DIERS) was established within the AIChE in 1976 to develop methods for the design of emergency relief systems to handle runaway reactions. By the mid to late 1970s, process safety was a recognized technical specialty. The American Institute of Chemical Engineers (AIChE) formed the Safety and Health Division in 1979.

    In 1976, a runaway reaction occurred near Seveso, Italy that resulted in the release of 2,3,7,8-tetrachlorodibenzo-p-dioxin, commonly known as dioxin, into residential areas. Dioxin is a toxic chemical. Many people developed Chloracne, a skin disease, and a 17 km^2 (6.6 mile^2) area was made uninhabitable. This incident eventually led to stricter regulations for the process industries in the European Economic Community in 1982, under what is known as the Seveso Directive.

> **Nitroglycerine.** The manufacture of nitroglycerine is an example of how process safety has evolved. Alfred Nobel began manufacturing nitroglycerine in 1864. The process was to add fuming nitric acid and sulfuric acid to glycerin while keeping the temperature at about 20-25 °C (68-77 °F). Keeping the temperature under control was critical to the safety of the process. Figure 2.1 shows a picture of an early nitroglycerine process, in which an operator is charged with observing the temperature and stopping feeds to the reactor if the temperature got too high. Note that he is sitting on a one legged stool. The reasoning was that, if he fell asleep (watching a temperature indicator for 8 to 10 hours a day is not the most interesting thing to do), he would fall and wake up. Also, note the size of the reactor. This represents a huge amount of nitroglycerine in one place, so if it did explode, the damage would be considerable. Explosions were not uncommon. Alfred Nobel's own brother was killed in such an explosion.
>
> Enough explosions occurred that some locations banned the use of nitroglycerine. Alfred Nobel's major breakthrough was discovering that when mixed with an inert carrier, it became safer to handle. This form of nitroglycerine was called dynamite.
>
> The process for making nitroglycerine evolved from the large, original batch reactors to inherently safer small, continuous reactors (see Figure 2.2) that are a fraction of the size of the original reactors. This makes the reaction easier to control, because the heat removal capability is better (notice the cooling coils for the reactor in Figure 2.2) and the mixing intensity of the smaller reactor is higher. It also reduces the extent of the damage if an explosion did occur. Automated controls mean that no one has to stand in front of the reactor anymore

In 1984 there was a defining moment in the chemical industry. A release of methylisocyanate, a toxic and flammable material, occurred at a chemical plant in Bhopal, India. More than 3,000 people died. (This incident will be described in more detail in Section 3.15.) The chemical plant itself was originally well designed, and had many safeguards against this event. These were not maintained, however, and at the time of the incident were not functioning. The Bhopal event drove home the point that technical expertise alone was not enough and that hazards or risk management was as important as the technical aspect of process safety. The Bhopal incident led to the formation of the Center for Chemical Process Safety (CCPS) in 1985. The CCPS is a not-for-profit organization that is

PROCESS SAFETY BASICS 7

Figure 2.1. Picture of a nitroglycerine reactor in the 19th century.
"Alfred Nobel in Scotland". Nobelprize.org. Nobel Media AB 2014. Web. 15 Sep 2015.
<http://www.nobelprize.org/alfred_nobel/biographical/articles/dolan/>

Figure 2.2. Continuous nitroglycerine reactor, courtesy Biazzi SA (www.Biazzi.com).

part of the U.S. based American Institute for Chemical Engineers with a mission to improve industrial process safety.

Process safety management. A *management system* is a formally established and documented set of activities and procedures designed to produce specific results in a consistent manner on a sustainable basis. Process Safety Management (PSM), therefore, is a management system that is focused on prevention of, preparedness for, mitigation of, response to, or restoration from releases of chemicals or energy from a process associated with a facility. In this book PSM refers to the CCPS use of Risk Based Process Safety Management Systems as described in the balance of this chapter, as opposed to the OSHA PSM system (described below).

In 1985 the Chemical Manufacturers Association (CMA), which later became the American Chemical Council (ACC), issued PSM guidelines (Ref. 2.1). By 1989, the CCPS introduced a set of 12 process safety management elements (Ref. 2.2). The American Petroleum Institute (API) also issued PSM guidelines in 1990 (Ref. 2.3). The CCPS studied the various approaches at the time and gleaned the 12 characteristics from interactions with its member companies and traditional business process consulting firms that had significant experience in evaluating management systems. Those guidelines were the first generic set of principles to be compiled for use in designing and evaluating process safety management systems.

In 1992 the Occupational Safety and Health Administration (OSHA) issued the Process Safety Management of Highly Hazardous Chemicals (OSHA PSM) regulation, which had its own, although similar, set of process safety management elements. The Environmental Protection Agency (EPA) issued its own version in 1995 under the authority of the Clean Air Act. This regulation is commonly referred to as RMP, or risk management plan, since the regulation requires the development and submittal of a risk plan based on the regulatory definitions and requirements.

OSHA PSM regulations cover the impacts of hazards to workers on-site; EPA RMP covers the impacts of hazards off-site. Both regulations are triggered by having more than a specified amount, commonly referred to as a Threshold Quantity (TQ), of specified chemicals, usually called Highly Hazardous Chemicals, or by more than 10,000 pounds of a flammable material in one location. In this sense, PSM as defined by the CCPS is much broader than OSHA PSM and EPA RMP.

Process safety practices and formal safety management systems have been in place in some companies for many years. PSM is widely credited for reductions in

PROCESS SAFETY BASICS

major accident risk and in improved process safety performance in the process industry. Nevertheless, many organizations continue to be challenged by inadequate management system performance, resource pressures, and stagnant process safety results. To promote PSM excellence and continuous improvement throughout the process industries, the CCPS created *risk-based process safety* (RBPS) as the framework for the next generation of process safety management (Ref. 2.4). The RBPS elements, along with the corresponding OSHA PSM / EPA RMP elements are shown in Table 2.1.

Risk: To discuss a risk-based process safety program, the concept of risk must be understood. A typical dictionary definition of risk, for example the Mirriam-Webster on-line dictionary, is "the possibility of loss or injury" or "someone or something that creates or suggests a hazard". The CCPS definition has three elements as opposed to two. They are: the hazard (what can go wrong), the magnitude (how bad can it be) and the likelihood (how often can it happen). Thus, in the process industries, understanding the risk associated with an activity requires answering the following questions:

1. What can go wrong? (human injury, environmental damage, or economic loss).
2. How bad could it be? (magnitude of the loss or injury).
3. How often might it happen? (likelihood of the loss or injury).

Figure 2.3 is an illustration of this. What can go wrong? The ducklings can fall through the grate. How bad can it be? One or more of them can fall through and be lost. How often might it happen? In this case, given the size of the ducklings, it probably happens every time they try to cross the grate.

Not all processes have the same amount of risk. Understanding risk helps a company decide how to shape its process safety management activities.

Resources, that is, money and people, are finite. When you design and run a facility, you can select from a wide range of options in deciding how much technical rigor to incorporate into the process safety management activities at your organization's facility (with the minimum requirement being complying with local and federal regulations). In other words, a process with low risk does not need the same amount of rigor in the application of the process safety elements as one with high risk.

Table 2.1. Comparison of RBPS elements to OSHA PSM elements.

CCPS RBPS Element	OSHA PSM/EPA RMP Elements
Commit to Process Safety	
1. Process Safety Culture	
2. Compliance with Standards	Process Safety Information
3. Process Safety Competency	
4. Workforce Involvement	Employee Participation
5. Stakeholder Outreach	Stakeholder Outreach (EPA RMP)
Understand Hazards and Risk	
6. Process Knowledge Management	Process Safety Information
7. Hazard Identification and Risk Analysis	Process Hazard Analysis
Manage Risk	
8. Operating Procedures	Operating Procedures
9. Safe Work Practices	Operating Procedures Hot Work Permits
10. Asset Integrity and Reliability	Mechanical Integrity
11. Contractor Management	Contractors
12. Training and Performance Assurance	Training
13. Management of Change	Management of Change
14. Operational Readiness	Pre-startup Safety Review
15. Conduct of Operations	
16. Emergency Management	Emergency Planning and Response
Learn from Experience	
17. Incident Investigation	Incident Investigation
18. Measurement and Metrics	
19. Auditing	Compliance Audits
20. Management Review and Continuous Improvement	

PROCESS SAFETY BASICS 11

Figure 2.3. Illustration of risk.

Commitment to process safety is the cornerstone of process safety excellence. Organizations generally do not improve without strong leadership and solid management commitment.

For process safety, management needs to recognize that process safety is not the same as personal safety, and move beyond personal safety programs. You will learn if the management of your company is committed to process safety by its actions.

Organizations that **understand hazards and risk** are better able to allocate limited resources in the most effective manner. Industrial experience has demonstrated that businesses using hazard and risk information to plan, develop, and deploy stable, lower-risk operations are much more likely to enjoy long-term success.

Managing risk focuses on four issues: (1) prudently operating and maintaining processes that pose the risk, (2) managing changes to those processes to ensure that the risk remains tolerable, (3) maintaining the integrity of equipment and assuring quality of materials, fabrications, and repairs, and (4) preparing for, responding to, and managing incidents that do occur. Managing risk helps a company or a facility deploy management systems that help sustain long-term, incident-free, and profitable operations.

Learning from experience involves monitoring, and acting on, internal and external sources of information. Despite a company's best efforts, operations do not always proceed as planned, accidents and near misses occur. A *near miss* is an event in which an accident (that is, property damage, environmental impact, or human loss) or an operational interruption could have plausibly resulted if circumstances had been slightly different. Organizations must be ready to turn their mistakes – and those of others – into opportunities to improve process safety efforts.

The twenty elements are described in more detail in the following sections. As a new engineer or someone new to process safety, some of these elements will have a more direct impact on you than others, but all have some impact. For example, learning about the Codes and Standards that affect your process and location will be an important part of your first few years in industry, whereas you may not be involved in Stakeholder Outreach at all. Nevertheless, the effort expended by the organization on stakeholder outreach may have a direct impact on how you will have to approach process safety at your locale.

Pillar: Commit to Process Safety

There are five elements to the pillar of Commit to Process Safety

- Process Safety Culture (Section 2.2)
- Compliance with Standards (Section 2.3)
- Process Safety Competency (Section 2.4)
- Workforce Involvement (Section 2.5)
- Stakeholder Outreach (Section 2.6)

2.2 Process Safety Culture

Case Study. On January 28, 1986, the space shuttle Challenger exploded 73 seconds after liftoff from Kennedy Space Center, killing all seven astronauts aboard (Figure 2.4). A field assembly joint in the right-hand solid rocket booster had failed, leaking hot combustion gases which, in turn, breached the liquid hydrogen vessel in the shuttle's external fuel tank assembly. The associated liquid oxygen vessel failed shortly thereafter, and the resulting catastrophic explosion destroyed the shuttle.

Figure 2.4. Challenger Disaster, courtesy NASA.

A subsequent investigation by a presidential commission revealed significant weaknesses in NASA's safety culture, which had set the stage for this disaster. These weaknesses included: 1) the tolerance of a situation in which production pressures – in this instance, the emphasis on maintaining an aggressive launch schedule – overshadowed safety concerns, 2) the gradual acceptance of increasing levels of damage to the field joints, as determined from post-launch inspections, as being a normal occurrence, even though this was in violation of design specifications and established safety requirements, 3) a can-do attitude, based upon past successes, that limited NASA's sense of vulnerability, and 4) a hierarchical structure and attitude that limited both the free exchange of information (especially disparate opinions) and the credibility given to the technical experts who were lower in the NASA structure or in contractor organizations.

Overview. Process safety culture has been defined as, "the combination of group values and behaviors that determine the manner in which process safety is managed" (Ref. 2.2). More succinct definitions include, "How we do things around here", "What we expect here" and "How we behave when no one is watching."

Investigations of catastrophic events have identified common process safety culture weaknesses that are often factors in other serious incidents. While the example of the Challenger explosion is not from the chemical or petrochemical industry, it illustrates the importance of a process safety culture, as well as illustrating that process safety management systems can apply to other endeavors.

The following features will help a company achieve a good process safety culture:

Maintain a sense of vulnerability. The organization maintains a high awareness of process hazards and their potential consequences and is constantly vigilant for indications of system weaknesses that might foreshadow more significant safety events.

Empower individuals to successfully fulfill their safety responsibilities. The organization provides clear delegation of, and accountability for, safety-related responsibilities. Accordingly, employees are provided the necessary authority and resources to allow success in their assigned roles. Personnel accept and fulfill their individual process safety responsibilities, and management expects and encourages the sharing of process safety concerns by all members of the organization.

Defers to expertise. The organization places a high value on the training and development of individuals and groups. The authority for key process safety decisions naturally migrates to the proper people based upon their knowledge and expertise, rather than their rank or position.

Ensures open and effective communications. Healthy communication channels exist both vertically and horizontally within the organization. Vertical communications are two way – managers listen as well as speak. Horizontal communications ensure that all workers have the information. The organization emphasizes promptly observing and reporting non-standard conditions to permit the timely detection of weak signals that might foretell safety issues.

Establishes a questioning/learning environment. The organization strives to enhance risk awareness and understanding as a means to continuous improvement in process safety performance. Enhancements are implemented in various ways, including:

- Performing appropriate and timely risk assessments.
- Promptly and thoroughly investigating incidents.
- Looking beyond the facility or company for applicable lessons
- Sharing and applying lessons learned throughout the organization, as appropriate.

The organization recognizes that catastrophic events typically have complex causes; consequently, overly simple solutions are avoided when addressing process safety issues. For example, the response to a release caused by pipe corrosion does not stop with replacing the pipe. Was the corrosion rate excessive? If so, why? Does this suggest a processing problem that must be addressed? Can a similar process plant within the organization be vulnerable to the same issue? Was the inspection frequency appropriate for the known corrosion rate? If not, why?

Fosters mutual trust. Employees trust managers to do the right thing in support of process safety. Managers trust employees to shoulder their share of responsibility for performance and to report potential problems and concerns promptly. However, even though mutual trust exists, people are willing to accept others evaluating or checking their actions related to critical tasks/activities that impact process safety risks.

Provides timely response to process safety issues and concerns. Priorities are placed on the timely communication and response to lessons learned from incident investigations, audits, risk assessments, and so forth. Mismatches between practices and procedures (or standards) are resolved in a timely manner to prevent normalization of deviance. The organization emphasizes the timely reporting and resolution of employee concerns.

PROCESS SAFETY BASICS 15

An AIChE presentation about the Challenger Case History can be found at: http://www.aiche.org/ccps/topics/elements-process-safety/commitment-process-safety/process-safety-culture/challenger-case-history.

2.3 Compliance with Standards

Case Study. On February 20, 2003, a dust explosion and fire damaged a manufacturing facility in Corbin, Kentucky, fatally injuring seven workers. The facility produced fiberglass insulation for the automotive industry. Investigators found that the explosion was fueled by resin dust that had accumulated in a production area, which was likely ignited by flames from a malfunctioning oven. The resin involved was a phenolic binder used in producing fiberglass mats. The investigation also determined that the company did not fully understand the risks of the combustible resin dust and did not follow industry consensus standards for preventing combustible dust explosions. These technologies are listed in numerous industry standards and guidelines (Ref. 2.5)

Overview: *Compliance with Standards.* A system to identify, develop, acquire, evaluate, disseminate, and provide access to applicable standards, codes, regulations, and laws that affect process safety. The *standards* system addresses both internal and external standards; national and international codes and standards, industry association guidance and practices; and local, state, and federal regulations and laws. The system makes this information easily and quickly accessible to potential users. The *standards* system interacts in some fashion with every RBPS management system element. Knowledge of and conformance to *standards* helps a company: 1) operate and maintain a safe facility, 2) consistently implement process safety practices, and 3) minimize legal liability. The s*tandards* system also forms the basis for the standards of care used in an audit program to determine management system conformance. The *standards* system provides a communication mechanism for informing management and personnel about the company's obligations and compliance status.

Table 2.2 provides some examples of the types of process safety obligations addressed by this element. *Note: This list is for example only and does not represent a complete list.* In Table 2.1 you will see that *standards* fall under the Process Safety Information element of the OSHA PSM and EPA RMP regulations. These codes, industry standards and recommended practices represent consensus on the approved way to perform certain engineering activities. A few examples are vessel design and fabrication, inspections, relief valve sizing, and facility siting.

In the US, OSHA and EPA use these codes as a basis for Recognized and Generally Accepted Good Engineering Practices (RAGAGEP) and expect that

Table 2.2. Examples and sources of process safety related standards, codes, regulations, and laws.

Voluntary Industry Standards & Consensus Codes
American Petroleum Institute Recommended Practices (Ref. 2.6)
American Chemistry Council Responsible Care® Management System and RC 12001 (Ref. 2.7)
ISO 12001 – Environmental Management System (Ref. 2.8)
OHSAS 18001 – International Occupational Health and Safety Management System (Ref. 2.9)
Organization for Economic Cooperation and Development – Guiding Principles on Chemical Accident Prevention, Preparedness, and1 Response, 2003 (Ref. 2.10)
American National Standards Institute (Ref. 2.10)
American Society of Mechanical Engineers (Ref. 2.12)
The Chlorine Institute (Ref. 2.13)
The Instrumentation, Systems, and Automation Society (Ref. 2.14)
National Fire Protection Association (Ref. 2.15)

U.S. Federal, State, and Local Laws and Regulations
U.S. OSHA – Process Safety Management Standard (29 CFR 1910.119) (Ref. 2.16)
U.S. Occupational Safety and Health Act – General Duty Clause, Section 5(a)(1) (Ref. 2.17)
U.S. EPA – Risk Management Program Regulation (20 CFR 68) (Ref. 2.18)
Clean Air Act – General Duty Requirements, Section 112(r)(1) (Ref. 2.19)
California Risk Management and Prevention Program (Ref. 2.20)
New Jersey Toxic Catastrophe Prevention Act (Ref. 2.21)
Contra Costa County Industrial Safety Ordinance (Ref. 2.22)
Delaware Extremely Hazardous Substances Risk Management Act (Ref. 2.23)
Nevada Chemical Accident Prevention Program (Ref. 2.24)

International Laws and Regulations
Australian National Standard for the Control of Major Hazard Facilities (Ref. 2.25)
Canadian Environmental Protection Agency – Environmental Emergency Planning, CEPA, 1999 (Section 200) (Ref. 2.26)
European Commission Seveso II Directive (Ref. 2.27)
Korean OSHA PSM Standard (Ref. 2.28)
Malaysia – Department of Occupational Safety and Health (DOSH) Ministry of Human Resources Malaysia, Section 16 of Act 512 (Ref. 2.29)
Mexican Integral Security and Environmental Management System (SIASPA) (Ref. 2.30)
United Kingdom, Health and Safety Executive COMAH Regulations (Ref. 2.31)

companies will comply with them. The same will be true of many local regulatory agencies and of those in other countries. Appendix A provides an example of list of codes a company may choose to use as RAGAGEP.

PROCESS SAFETY BASICS 17

Companies can have different ways of ensuring compliance with standards. Some may develop their own internal standards that are based on applicable consensus codes and standards, such as listed in Table 2.2. Others may rely on Subject Matter Experts (SME) who review projects for code compliance, or some combination of both, or another method entirely. As a new engineer in a company, you will have to learn how your company ensures compliance with standards.

2.4 Process Safety Competency

Case Study. On April 21, 1995, a blender containing a mixture of sodium hydrosulfite, aluminum powder, potassium carbonate, and benzaldehyde exploded and triggered a major fire at a specialty chemical plant in Lodi, New Jersey. Five employees were killed and many more were injured. Most of the plant was destroyed as a result of the fire, and other nearby businesses were destroyed or significantly damaged. Property damage at the plant was estimated at $20M. An EPA/OSHA joint chemical accident investigation team determined that the immediate cause of the explosion was the inadvertent introduction of water and heat into water-reactive materials during the mixing operation (Ref 2.32). During an emergency operation to offload the blender contents, the material ignited and a deflagration occurred. The investigation team identified the following root causes and contributing factors of the accident:

- An inadequate process hazards analysis was conducted, and appropriate preventive actions were not taken.
- Standard operating procedures and training were inadequate.
- The decision to re-enter the plant and offload the blender was based on inadequate information.
- The blender used was inappropriate for the materials blended.
- Communications between the plant and the company that provided the blending technology was inadequate; the use for mixing water reactive materials not recommended by the equipment manufacturer.
- The training of fire brigade members and emergency responders was inadequate.

Overview: Developing and maintaining *process safety competency* encompasses three interrelated actions: 1) continuously improving knowledge and competency, 2) ensuring that appropriate information is available to people who need it, and 3) consistently applying what has been learned.

Process safety competency is harder to assess than, for example, whether applicable codes are being followed. New engineers should be looking for whether they and other people are provided with training opportunities, have a chance to apply their training and if they have access to the necessary process safety information.

2.5 Workforce Involvement

Example. During a hazard identification review you or another engineer is called upon to describe in detail how some aspect of a process or piece of equipment operates – only to have a seasoned operator respond, "That's all well and good, but let me tell you what *really* happens on a Sunday at 3:00 a.m."

Overview. Personnel, at all levels and in all positions in an organization, should have roles and responsibilities for enhancing and ensuring the safety of the organization's operations. However, some workers may not be aware of all of their opportunities to contribute. Some organizations may not effectively tap into the full expertise of their workers or, worse, may even discourage workers who seek to contribute through what the organization views as a nontraditional role. *Workforce involvement* provides a system for enabling the active participation of company and contract workers in the design, development, implementation, and continuous improvement of the RBPS management system.

Those personnel directly involved in operating and maintaining the process are those most exposed to the hazards of the process. The workforce involvement element provides an equitable mechanism for workers to be directly involved in protecting their own welfare. Furthermore, these workers are potentially the most knowledgeable people with respect to the day-to-day details of operating the process and maintaining the equipment and facilities, and may be the sole source for some types of knowledge gained through their unique experiences. Workforce involvement provides management a formalized mechanism for tapping into this valuable expertise.

In Table 2.1 workforce involvement corresponds to a requirement for Employee Participation in the OSHA PSM and EPA RMP regulations. Specifically, this is a requirement for the presence of people with "experience and knowledge specific to the process being evaluated" at a hazard identification study of a covered process. This is because these agencies recognize the need to have a person who actually knows the process to be involved. Proactive companies will expand that to include workers directly involved in maintenance and operations at these reviews and will encourage their honest input. Subject matter experts such as process engineers, mechanical engineers, material engineers, etc. should be relied on for technical information and other process safety information. Operators and mechanics should be relied on for evaluating the understanding of the process, the clarity and efficiency of procedures, and an understanding of what is being done in the field vs. what engineering and management think is being done. Other localities and countries are also likely to have regulatory requirements on workforce involvement.

This proactive engagement would illustrate at least two positive things, the right people are involved in the review, and the workforce, down to the operating staff, feels free to provide candid views without fear of adverse consequences.

2.6 Stakeholder Outreach

Example. In the late 1990s, as a result of regulations from the U.S. Environmental Protection Agency (EPA), many facilities handling hazardous materials that could affect the public were required to undertake risk communication efforts with their communities to comply with the EPA's Risk Management Plan (RMP) rules (Ref. 2.18). In highly populated areas having a large industrial base, many companies collaborated by having regional RMP communication events. In some cases, multi-year programs were created to: 1) engage stakeholder groups, 2) identify their concerns and needs, 3) develop communication/outreach plans, 4) conduct planned activities, and 5) follow up on the results. One of the largest such activities was conducted in the Houston, Texas, area where more than 120 facilities coordinated their RMP outreach activities over a 4-year period. Those activities helped nurture relationships with communities, regulators, local emergency response agencies, and nongovernmental community groups.

Effective stakeholder relations can be beneficial to process safety and plant operations in general. As an example, in 1998 a specialty gas repackaging firm serving the semiconductor industry in California applied for a permit to expand its facility. Despite the opposition of a national environmental group, the company's long-term effective outreach program had fostered overwhelming local support, and the permit was granted.

Overview. *Stakeholder outreach* is a process for: 1) seeking out individuals or organizations that can be or believe they can be affected by company operations and engaging them in a dialogue about process safety, 2) establishing a relationship with community organizations, other companies and professional groups, and local, state, and federal authorities, and 3) providing accurate information about the company and facility's products, processes, plans, hazards, and risks. This process ensures that management makes relevant process safety information available to a variety of organizations. This element also encourages the sharing of relevant information and lessons learned with similar facilities within the company and with other companies in the industry group. A key aspect of the mission of the AIChE CCPS is to provide forums for sharing of best practices and information concerning hazards and failures of various process safeguards. Finally, the *outreach* element promotes involvement of the facility in the local community and facilitates communication of information and facility activities that could affect the community.

Pillar: Understand Hazards and Risks

There are two elements to the pillar Understand Hazards and Risk

- Process Knowledge Management (Section 2.7)
- Hazard Identification and Risk Analysis (Section 2.8)

2.7 Process Knowledge Management

Case Study. On February 19, 1999, an explosion in Lehigh County, Pennsylvania, resulted in 5 fatalities, 14 injuries, and damage to several nearby buildings (Figure 2.5). The explosion occurred during manufacture of the facility's first production scale lot of hydroxylamine. According to the Chemical Safety and Hazard Investigation Board's (CSB's) investigation report (Ref. 2.32):

Figure 2.5. Building damage and charge tank crater, Hydroxylamine explosion, courtesy CSB.

"The incident demonstrates the need for effective process safety management and engineering throughout the development, design, construction, and startup of a hazardous chemical production process . . . [D]efficiencies in "process knowledge and documentation" and "process safety reviews for capital projects" significantly contributed to the incident."

The CSB report goes on to say:

"[T]he development, understanding, and application of process safety information during process design was inadequate for managing the explosive decomposition hazard of hydroxylamine. During pilot-plant operation, management became aware of the fire and explosion hazards of hydroxylamine concentrations in excess of 70-wt. %, as documented in the MSDS.

This knowledge was not adequately translated into the process design, operating procedures, mitigative measures, or precautionary instructions for process operators. [The] hydroxylamine production process, as designed, concentrated hydroxylamine in a liquid solution to a level in excess of 85 wt. %. This concentration is significantly higher than the MSDS referenced 70 wt. %%concentration at which an explosive hazard exists."

The CSB investigators concluded that lack of process knowledge and failure to properly apply the available process knowledge led directly to the explosion. This incident is discussed in more detail in Chapter 3.

Overview. Understanding risk depends on accurate process knowledge. Thus, this element underpins the entire concept of risk-based process safety management; the RBPS methodology cannot be efficiently applied without an understanding of risk. The *process knowledge* element primarily focuses on information that can be recorded in documents, such as:

- Written technical documents and specifications.
- Engineering drawings and calculations.
- Specifications for design, fabrication, and installation of process equipment.
- Selection of safe operating limits for pressure, temperature, level, concentration, etc.
- Other written documents such as material safety data sheets (MSDSs).

The term *process knowledge* will be used to refer to this collection of information. The *knowledge* element involves work activities associated with compiling, cataloging, and making available a specific set of data that is normally recorded in paper or electronic format. However, *knowledge* implies understanding, not simply compiling data. In that respect, the *competency* element complements the *knowledge* element in that it helps ensure that users can properly interpret and understand the information that is collected as part of this element.

Development and documentation of process knowledge starts early and continues throughout the life cycle of the process. For example, early laboratory efforts to develop new materials, characterize these materials, and evaluate the synthesis route (including the potential for runaway reaction or other inherent hazards) normally become part of the process knowledge. Efforts continue through the design, hazard review, construction, commissioning, and operational phases of the life cycle. Many facilities place special emphasis on reviewing process knowledge for accuracy and thoroughness immediately prior to conducting a risk analysis or management of change review.

Process knowledge is typically collected and cataloged as hard copy documents stored in file cabinets or libraries and electronic files or databases maintained on computer networks. When you begin working at an organization, you will be able to see how process knowledge is kept and should familiarize yourself with it.

In Table 2.1 you will see that process knowledge management corresponds to the Process Safety Information (PSI) element of OSHA PSM and EPA RMP regulations, which requires a written compilation of PSI. These regulations specifically call for written information on the hazards of the chemicals used or produced, and on the technology and equipment used in covered processes. Again, local authorities and other countries are likely to have similar requirements.

2.8 Hazard Identification and Risk Analysis

Case Study. In 1998, a major explosion and fire occurred at the Longford gas-processing facility in Victoria, Australia (Ref. 2.33). Two employees were killed and eight others were injured. The incident caused the destruction of one gas separation plant and the shutdown of two others at the facility. This disrupted gas supplies across the state for two weeks, resulting in 250,000 workers being sent home as factories and businesses were forced to shut down. The incident as workers attempted to recover from a process upset that had embrittled the metal of a heat exchanger. Had a thorough risk study been conducted it could have identified the potential for a loss of lean oil to create dangerously low temperatures in the process equipment. A risk study had been planned three years prior to the accident, but had not been done.

Overview. *Hazard Identification and Risk Analysis* (HIRA) is a collective term that encompasses all activities involved in identifying hazards and evaluating risk at facilities, throughout their life cycle, to make certain that risks to employees, the public, or the environment are consistently controlled within the organization's risk tolerance. Note: There are many different terms to describe the tasks that comprise an HIRA, as shown in Table 2.3. In fact, you will probably not be asked to participate in an "HIRA", even though you most likely will participate in various types of hazards identification and risk analysis studies. In some companies, the term Process Hazard Analysis (PHA) has come to specifically mean the study done to comply with the OSHA PSM standard, and another term, such as Hazard Assessment is used to mean any other hazard identification study.

To manage risk, hazards must first be identified, and then the risks should be evaluated and determined to be tolerable or not. The risk understanding developed from these studies forms the basis for establishing many of the other process safety management activities undertaken by the facility. An incorrect perception of risk at any point could lead to either inefficient use of limited resources or unknowing acceptance of risks exceeding the true tolerance of the company or the community.

These studies typically address three main risk questions to a level of detail commensurate with analysis objectives, life cycle stage, available information, and resources. The three main risk questions are:

1. Hazard – What can go wrong?
2. Consequences – How bad could it be?
3. Likelihood – How often might it happen?

To do a hazard assessment, a review team questions process experts about possible hazards and judges the risk of any hazards that are identified. A suite of tools is available to accommodate varying analysis needs. Tools for hazard identification or qualitative risk analysis include hazard and operability analysis (HAZOP), what-if/checklist analysis, and failure modes and effects analysis (FMEA). Tools for semi-quantitative risk analysis include failure modes, effects, and criticality analysis (FMECA) and layer of protection analysis (LOPA). Tools for more detailed quantitative risk analysis include fault trees, event trees and consequence modeling. The following books cover these topics:

- *Guidelines for Hazard Evaluation Procedures*, Third Edition with Worked Examples, Ref. 2.34.

Table 2.3 Hazard evaluation synonyms

Process hazard(s) analysis	Predictive hazard evaluation	Hazard assessment
Process hazard(s) review	Process risk review	Hazard and risk analysis
Process safety review	Process risk survey	Hazard study

- *Layer of Protection Analysis – Simplified Process Risk Analysis*, Ref. 2.35
- *Guidelines for Chemical Process Quantitative Risk Analysis*, Second Edition, Ref. 2.36.
- *Guidelines for Initiating Events and Independent Protection Layers*, Ref. 2.37.

The results of the review process are typically documented in a worksheet form which varies in detail, depending on the stage of the project and the evaluation method used. Table 2.4 shows a typical HAZOP analysis worksheet.

HIRA encompasses the entire spectrum of risk analyses, from simple and qualitative to detailed and quantitative. Some companies may judge the existence of a risk of explosion to be an unacceptable risk, regardless of its likelihood. Others may be unwilling to accept an explosion risk unless it can be shown that the expected frequency of explosions is less than a specified likelihood, such as one in a million per year. In such cases a simple or detailed risk analysis is necessary.

Risk studies on operating processes are typically updated or revalidated on a regular basis. Various countries and localities may have specific requirements concerning HIRA. A process hazard analysis (PHA) is an HIRA that meets the specific regulatory requirements of the OSHA PSM standard in the U.S. (Ref. 2.16).

As a new engineer, or an engineer new to process safety, it is very likely that you will participate in some form of HIRA in your first few years in the process industries In Table 2.1 the HIRA element corresponds to the Process Hazard Analysis (PHA) requirement of the OSHA PSM and EPA RMP regulations. These regulations require that a PHA be performed on a covered process, and that it be updated every 5 years. Local authorities and other countries are likely to have similar requirements.

Table 2.4 Typical HAZOP review table format.

Area:			Meeting Date:	
Drawing Numbers:			Team Members:	
Deviation	Causes	Consequences	Safeguards	Recommendations
No Flow	Valve fails closed			
	Plug in line			

Pillar: Manage Risk

There are nine elements to the pillar of Manage Risk.

- Operating Procedures (Section 2.9)
- Safe Work Practices (Section 2.10)
- Asset Integrity and Reliability (Section 2.11)
- Contractor Management (Section 2.12)
- Training and Performance (Section 2.13)
- Management of Change (Section 2.14)
- Operational Readiness (Section 2.15)
- Conduct of Operations (Section 2.16)
- Emergency Management (Section 2.17)

2.9 Operating Procedures

Case Study. On December 13, 1994, an explosion occurred shortly before restart of an ammonium nitrate plant in Port Neal, Iowa. Four employees were killed, and an additional 18 employees were admitted to the hospital. Public evacuations were ordered up to 15 miles from the facility, and property damage was estimated at $120 million. The plant had been temporarily shut down 18 hours earlier and operators were preparing to restart the plant at the time of the explosion. The Environmental Protection Agency's (EPA's) Chemical Accident Investigation Team concluded that the explosion resulted from the lack of written procedures for conducting a temporary shutdown on the ammonium nitrate plant (Ref. 2.38). More details concerning this incident are presented Section 3.6.

Overview. *Operating procedures* are written instructions (including procedures that are stored electronically and printed on demand) that list the steps for a given task, and describe the manner in which the steps are to be performed. Good procedures also describe the process, hazards, tools, protective equipment, and controls in sufficient detail that operators understand the hazards, can verify that controls are in place, and can confirm that the process responds in an expected manner. Procedures also provide instructions for troubleshooting when the system does not respond as expected. Procedures should specify when an emergency shutdown should be executed and should also address special situations, such as temporary operation with a specific equipment item out of service. Operating procedures are normally used to control activities such as transitions between products, periodic cleaning of process equipment, preparing equipment for certain maintenance activities, and other activities routinely performed by operators. The scope of this element includes those operating procedures that describe the tasks required to safely start up, operate, and shut down processes, including emergency shutdown.

A consistent high level of human performance is a critical aspect of any process safety program. Indeed, a less than adequate level of human performance will adversely impact all aspects of operations. Without written procedures, a facility can have no expectation that the intended procedures and methods are used by each operator, or even that an individual operator will consistently execute a particular task in the intended manner.

Procedure writing is something that a new engineer may become involved with. Procedures are often jointly developed by operators and process engineers who have a high degree of involvement and knowledge of process operations. (The term operator is used to describe the person who directly controls the process either via a control system or manipulation of field equipment; note that many facilities use other terms, such as technician, to describe this function.) Operators, supervisors, engineers, and managers are often involved in the review and approval of new procedures or changes to existing procedures. Other work groups, such as maintenance, should also be involved if the operating procedures could potentially affect them.

Once procedures have been developed and approved the procedures should be precisely followed, without exception, every time, or they need to be changed through a formal Management of Change business process as described in a subsequent section of this chapter.

In Table 2.1 you will see that Operating Procedures are an element of the OSHA PSM and EPA RMP regulations. These rules require that there be procedures for normal operations, startup and shutdown, emergency shutdown, and emergency operations, and that they be kept up-to-date. Local authorities and other countries are likely to have similar requirements.

2.10 Safe Work Practices

Case Study. On July 17, 2001, an explosion occurred at the Motiva Enterprises LLC Delaware City Refinery (DCR) in Delaware City, Delaware. A crew of maintenance contractors was repairing grating on a catwalk in a sulfuric acid (H_2SO_4) storage tank farm when a spark from their hot work ignited flammable vapors in one of the storage tanks. The tank separated from its floor, instantaneously releasing its contents (Figure 2.6). One contractor was killed, eight others were injured. Other tanks in the tank farm also released their contents. A fire burned for approximately one-half hour; and H_2SO_4 reached the Delaware River, resulting in significant damage to aquatic life. The tanks stored fresh and spent H_2SO_4 used in the refinery's sulfuric acid alkylation process. Spent acid normally contains small amounts of flammable material. Over the years, the tanks had experienced significant localized corrosion. Leaks were found on the shell of one tank annually from 1998 through May 2001. Motiva allowed hot work to be performed in the vicinity of a tank with holes in its roof and shell. Hot work should not have been authorized. In addition, once the work was authorized, inadequate precautions were taken to prevent an ignition of flammable vapors. (Ref. 2.39)

Overview. *Safe work practices* are a formalized process to help control hazards and manage risk associated with work done that is not directly involved with process operations. Maintenance and inspection activities within process areas are examples of work that would be managed with a safe work practice. Making and breaking connections to unload a railcar would likely be covered by an operating procedure, whereas breaking a connection to a pressure transmitter would be considered a non-routine work activity and included in the scope of the safe work practices *(safe work)* element. Safe work practices are not specific to a single activity or work instruction. They are generic and often written for use across a facility or the entire organization. Examples of safe work practices include lock-out-tag-out (LOTO), hot work, line break, confined space entry, etc. Many countries may have specific regulations for safe work practices which industries are required to follow.

Figure 2.6. Collapsed tank at Motiva refinery, courtesy CSB.

Often organization's occupational safety and health departments oversee and manage safe work procedures. Safe work procedures are essential to process safety. These procedures allow for the maintenance, inspection, and repair of process equipment, vessels, controls, and piping in a consistent and safe manner.

Safe work procedures are often supplemented with permits (i.e., a checklist that includes an authorization step). A more comprehensive list of safe work practices is provided in Table 2.5.

Some facilities also include procedures or practices that protect against standard industrial hazards, such as falling, in the scope of this element.

The OSHA PSM and EPA RMP regulations call for hot work procedures. Other safe work practices are often required by other regulations, regardless of the magnitude of chemical or other hazards present at a facility.

2.11 Asset Integrity and Reliability

Case Study. On August 6, 2012, the Chevron U.S.A. Inc. Refinery in Richmond, California (Chevron Richmond Refinery) experienced a catastrophic pipe rupture in the #4 Crude Unit (Figure 2.7). The ruptured pipe released a flammable hydrocarbon process fluid which then partially vaporized into a large vapor cloud that engulfed nineteen employees.

Table 2.5. Activities typically included in the scope of the safe work element

Safe work procedures to control general hazards or protect personnel from a hazard or hazardous environment include:
- Lockout/tagout and/or control of energy hazards.
- Line breaking/opening of process equipment.
- Confined space entry.
- Hot work authorization.
- Access to process areas by unauthorized personnel.
- Hot tapping lines and equipment.

Safe work procedures to protect against mishaps that could have catastrophic secondary effects include:
- Excavation in or around process areas.
- Operation of vehicles in process areas.
- Lifting over process equipment.
- Use of other heavy construction equipment in or around process areas.

Safe work procedures to control special hazards include:
- Use of explosives/blasting operations.
- Use of ionizing radiation (e.g., to produce x-ray images of process equipment).

Safe work procedures to prevent unauthorized impairment of safety systems include:
- Fire protection system impairment.
- Temporary isolation of relief devices.
- Temporary bypassing or jumpering of interlocks.

PROCESS SAFETY BASICS

Approximately two minutes after the release, the flammable portion of the vapor cloud ignited. Six Chevron employees suffered minor injuries during the incident and subsequent emergency response efforts. The flash fire could have resulted in much more serious consequences for the workers present at the scene. The fire burned for hours and resulted in a lengthy (many months) and costly shutdown of the process unit and outrage from thousands of local residents exposed to smoke who sought medical evaluations at local hospitals. The rupture occurred in a 52 inch long section of 8 inch diameter line on the side of a distillation column. Analysis found that the 52-inch component where the rupture occurred had experienced extreme thinning and had lost, on average, 90 percent of its original wall thickness in the area near the rupture.

The thinning was caused by sulfidation corrosion, which is a known phenomenon, and the American Petroleum Institute (API) has a Recommended Practice (RP) about it: 939-C *Guidelines for Avoiding Sulfidation (Sulfidic) Corrosion Failures in Oil Refineries*. The line involved had 24 Condition Monitoring Locations (CMLs), but there were no CMLs on the 52 inch section of the piping.

Figure 2.7. Rupture in 52-inch component of line, courtesy CSB.

This section was more susceptible to the sulfidation corrosion, because it contained lower silicon levels than other sections of the piping (for piping components lower silicon levels are known to be susceptible to high sulfidation corrosion rates). Relying on inspection data from high silicon components, the

dangerous condition of the piping was not known, and opportunities to replace it were missed. (Ref. 2.40)

Overview. *Asset integrity and reliability* is the RBPS element that helps ensure that equipment is properly designed, installed in accordance with specifications, and remains fit for use until it is retired. *Asset integrity* involves activities, such as inspections and tests, necessary to ensure that equipment will be suitable for its intended application throughout its life. Maintaining containment of hazardous materials and ensuring that safety systems such as pressure relief devices and Safety Controls Alarms and Interlocks (SCAI) work when needed are two of the primary responsibilities of any facility.

This element primarily involves inspections, tests, preventive maintenance, predictive maintenance, and repair activities that are performed by maintenance and contractor personnel at operating facilities. It also involves quality assurance processes (e.g. positive materials identification for piping and gasket materials), including procedures and training, that support these activities.

At an operating facility, the *asset integrity* element activities are an integral part of day-to-day operation involving operators, maintenance employees, inspectors, contractors, engineers, and others involved in designing, specifying, installing, operating, or maintaining equipment. Mechanical engineers, in particular, are heavily involved in the preventive maintenance for equipment items. Instrumentation and Electrical (I&E) engineers often become involved in testing and maintenance of instruments and control loops. Process engineers can contribute by looking for and reporting leaks, unusual noises or odors, or detecting other abnormal conditions.

This element is covered by the Mechanical Integrity requirement of the OSHA PSM and EPA RMP regulations (see Table 2.1). The regulation requires written procedures for inspection and testing, documentation that the inspection and testing was done and that deficiencies are corrected. Local authorities and other countries are likely to have similar requirements.

2.12 Contractor Management

Case Study. On the evening of July 6, 1988, an explosion and fire occurred on the Piper Alpha offshore platform, resulting in 167 fatalities and destruction of the platform. The incident occurred as the night crew was putting a condensate injection pump into service that, earlier in the day, had been taken out of service for maintenance. The night crew was aware of the maintenance activity and, in fact, had to authorize the electricians to close the switch at the motor control center so that the pump could be returned to service. The evening shift operators were likely told that all of the maintenance work scheduled to be performed by the maintenance group was either complete or had been deferred; however, they were not aware that other work being performed by a contractor was incomplete. In fact, the contract crew had removed a pressure safety valve on the pump discharge line for recertification and was unable to return the valve to service prior to 6 p.m. when they quit for the day. Once the pump was started, a large release of hydrocarbons ultimately led to the disaster. The subsequent investigation indicated that the contract supervisor responsible for the related maintenance job had not been properly trained in the safe work procedure. Furthermore, the investigation inferred that inadequacies in the emergency response training given (or in some cases, not given) to contractors on the oil platform likely contributed to the high loss of life in the accident (Ref. 2.40, 2.41).

Overview. Industry often relies upon contractors for everyday operations, certain specialized skills, and during periods of intense activity, such as maintenance turnarounds. These considerations, coupled with the potential lack of familiarity that contractor personnel may have with facility hazards and operations, pose unique challenges for the safe utilization of contract services. Contractor management is a system of controls to ensure that contracted services support both safe facility operations and the company's process safety and personal safety performance goals. This element addresses the selection, acquisition, use, and monitoring of such contracted services.

Companies are increasingly leveraging internal resources by contracting for a diverse range of services, including design and construction, maintenance, inspection and testing, and staff augmentation. In doing so, a company can achieve goals such as: 1) accessing specialized expertise that is not continuously or routinely required, 2) supplementing limited company resources during periods of unusual demand, and 3) providing staffing increases without the overhead costs of direct-hire employees.

However, using contractors brings an outside organization into the realm of the company's risk control activities. The use of contractors can place personnel who are unfamiliar with the facility's hazards and protective systems into locations where they could be affected by process hazards. Conversely, as a result of their work activities, the contractors may introduce new hazards to a facility, such as new chemicals or x-ray sources. Also, their activities onsite may unintentionally defeat or bypass facility safety controls. Companies need to recognize and address new challenges associated with using contractors. Tasks a company needs to do for contractor management include:

- Check contractors safety records when selecting them.
- Establish expectations, roles, and responsibilities for safety program implementation and performance.
- Ensure that contractor personnel are properly trained.
- Supply appropriate information to the contractor to ensure that the contractor can safely provide the contracted services.

Contractor management is also an element of the OSHA PSM and EPA RMP regulations. The aforementioned requirements are part of the regulation. Local authorities and other countries are likely to have similar requirements.

2.13 Training And Performance Assurance

Case Study. On January 4, 1966, a flash fire followed by a boiling liquid expanding vapor explosion (BLEVE) occurred within the liquefied petroleum gas (LPG) tank farm area of a refinery near Lyon, France. Eighteen people were killed and 81 were injured (Ref. 2.43). The liquefied gas storage facility was destroyed and the fire spread to nearby liquid hydrocarbon storage tanks. The incident was started when an attempt to drain water from the bottom of one of the LPG storage spheres. LPG flashing through the upstream drain valve (in a double block valve arrangement) that was being used to regulate the flow in the drain line caused both the upstream and downstream valves to freeze open, releasing LPG to the atmosphere through the open drain. If the water flow had been regulated with the downstream valve, the upstream valve would not have frozen open when LPG began to flash. This upstream valve, then, might have been used to stop the flow. This incident illustrates why the management system must ensure that everyone involved with a process is trained in the correct procedures and why their satisfactory performance must be periodically verified.

Overview. *Training* is practical instruction in job and task requirements and methods. It may be provided in a classroom or workplace, and its objective is to enable workers to meet some minimum initial performance standards, to maintain their proficiency, or to qualify them for promotion to a more demanding position. *Performance assurance* is the means by which personnel demonstrate that they have understood the training and can apply it in practical situations. Performance assurance is an ongoing process to ensure that workers meet performance standards and to identify where additional training is required.

Safe operation and maintenance of facilities that manufacture, store, or otherwise use hazardous chemicals requires qualified workers at all levels, from managers and engineers to operators and craftsmen. Training and other performance assurance activities are the basis for achieving high levels of human reliability. In this context, training broadly includes education in specific procedures governing operations, maintenance, safe work, and emergency planning and response, as well as in the overall process and its risks. Training takes place both in the workplace and the classroom, and it should be completed before a worker is allowed to work independently in a specific job position.

Training is an element of the OSHA PSM and EPA RMP regulations. The previous requirements are part of the regulation. Local authorities and other countries are likely to have similar requirements.

2.14 Management of Change

Case Study. The management of change (MOC) element has been called the minute-by-minute process risk assessment and control system. The significance of MOC – or the lack of it – was never more visible than in the 1974 Flixborough accident. This watershed event involved a temporary modification to piping between cyclohexane oxidation reactors. In an effort to maintain production, a temporary bypass line was installed around the fifth of a series of six reactors at a facility in Flixborough, England, in March 1974. This bypass failed while the plant was being restarted after unrelated repairs on June 1, 1974, releasing about 60,000 pounds of hot process material, composed mostly of cyclohexane. The resulting vapor cloud exploded, yielding an energy release equivalent to about 15 tons of TNT. The explosion completely destroyed the plant, and damaged nearby homes and businesses, killing 28, and injuring 89 employees and neighbors.

The temporary modification was constructed by staff whose expertise was insufficient design large pipes equipped with bellows – the design work for the change was treated more like pipefitting than large diameter, high pressure piping system design. As stated in the official report, "…they did not know that they did not know." An effective MOC system would have discovered the design flaw before the change was implemented, thus averting the disaster.

Overview. The Management of Change (MOC) element helps ensure that changes to a process do not inadvertently introduce new hazards or unknowingly increase risk of existing hazards. Many companies will define a change as anything that is not a *replacement-in-kind*. CCPS defines a change as *anything that changes the process safety information*. The purpose of management of change is to assess the risk associated with change, and mitigate that risk to acceptable levels. This is usually accomplished by a team using a defined methodology and approval process.

The MOC element includes a review and authorization process for evaluating proposed changes to facility design, operations, organization, or activities prior to implementation to make certain that no unforeseen new hazards are introduced and that the risk of existing hazards to employees, the public, or the environment is not unknowingly increased. It also includes steps to help ensure that potentially affected personnel are notified of the changes and pertinent documents, such as procedures, process safety knowledge, and so forth, are kept up–to-date.

If a proposed modification is made to a hazardous process without appropriate review, the risk of a process safety accident could increase significantly. MOC reviews are conventionally done in operating plants and increasingly done throughout the process life cycle at company offices that are involved with capital project design and planning. It is highly likely that, as a new engineer, you will become involved in MOC reviews in the first few years of your career.

A change request can come from anyone from an individual on the facility floor, to a project team, to a research or development team. One of the key things an individual in a facility needs to know is what constitutes a change. Qualified personnel, normally independent of the MOC originator, review the request to determine if any potentially adverse risk impacts could result from the change, and may suggest additional measures to manage risk. The nature of the review will usually depend on the extent of the change. Simple changes may only need a few knowledgeable people. A significant change may require a review that resembles a HIRA as discussed in Section 2.8. Based on the review, the change is either authorized for execution, amended, or rejected. Often, final approval for implementing the change comes from another designated individual, independent of the review team. A wide variety of personnel are normally involved in making the change, notifying or training potentially affected employees, and updating documents affected by the change.

MOC is specifically covered by OSHA PSM and EPA RMP regulations, and by regulations in other countries. The rule requires not only conducting the MOC review, but updating the Process Safety Information (PSI) and operating procedures affected and informing the workforce of the change.

Although the importance of management of change is understood by many organizations today, the importance of Management of Organizational Change (MOOC), also called Organizational Change Management (OCM), is less well understood. OCM covers changes in working condition, personnel changes, task allocation changes, organizational structure changes and policy changes. The CCPS *Guidelines for Managing Process Safety during Organizational Change* (Ref. 2.43) covers this subject in more detail.

2.15 Operational Readiness

Case Study. A refinery underwent a major turnaround of one of its units, which involved extensive lock-out-tag-out (LOTO) requirements for multiple pieces of equipment. Procedures were in place for LOTO, initial line breaking, and all other relevant safety processes. One unique aspect of this unit turnaround was that several pieces of equipment included in the process boundary were only operated on an occasional basis when feedstocks were changed and additional heating and/or cooling were needed.

All pre-startup checks were made per the established checklists, and the unit restarted smoothly. Unfortunately, six months later, when one of the occasionally used heat exchangers was put into service, a major leak of flammable gas ensued. The resulting fire caused a significant outage and equipment damage. It was discovered that the startup checks for this equipment had not been performed as a part of the overall turnaround, because this particular piece of equipment was not going to be used immediately. Although individual processes were in place to perform pre-startup checks, no overall operational readiness process had been established to ensure that the entire unit was ready to run.

Overview. The *operational readiness* element can be discussed from two perspectives. As the elements of process safety were first being formally identified, operational readiness involved confirming that new processes, modified processes, and changes to processes were safe to start up. This activity is referred to as Pre-Startup Safety Reviews (PSSRs). The advancement of risk based process safety has led many organizations to expand the scope of operational readiness to include verification that equipment and processes that have been shut down are in a safe condition for re-start.

Operational readiness should address startups from all types of shutdown conditions. Operational readiness also encompasses not only new and changed processes, but also the startup or restart of equipment and processes that have been opened up, inspected, repaired, etc. This can involve relatively small maintenance activities or it can follow large scale maintenance turnarounds involving many weeks of maintenance activities. Some processes may be shut down only briefly, while others may have undergone a lengthy maintenance/inspection/repair outage, or they may even have been mothballed for an extended period. Other processes may have been shut down for administrative reasons, such as a lack of product demand; for reasons unrelated to production at all; or as a precautionary measure, for example, because of an approaching storm. In addition to the shutdown duration, this element considers the type of work that may have been conducted on the process (e.g., possibly involving line-breaking) during the shutdown period to help focus the *readiness* review prior to startup.

Table 2.1 shows that the *readiness* element corresponds to the Pre-startup Safety Review in the OSHA PSM and EPA RMP regulations. In this case, however, operational readiness is defined more broadly than the OSHA process safety management pre-startup safety review element in that it specifically addresses startup from all shutdown conditions – not only those resulting from new or changed processes.

Experience has shown that the frequency of incidents is higher during process transitions such as startups. These incidents have often resulted from the physical process conditions not being exactly as they were intended for safe operation. Thus, it is important that the process status be verified as safe to start.

A review involves some or all of the following concerns and activities:

- Confirming that the construction and equipment of a process are in accordance with design specifications.
- Ensuring that adequate safety, operating, maintenance, and emergency procedures are in place.
- Ensuring that any safeguards that may have been bypassed during the outage are verified to be in service and operational.
- Confirming that all sensors, instruments, and valves are properly reset to the proper state or condition.
- Ensuring that training has been completed for all workers who may affect the process.

Also, for new processes, it confirms that an appropriate risk analysis has been performed, and that any recommendations have been resolved and implemented. Modified processes should have undergone a management of change (MOC) review. For all startups (including those after minor, short-term shutdowns not involving any changes), readiness reviews ensure that the process is safe to be released to operations by examining issues such as the equipment lineup, leak tightness, proper isolation from other systems not yet ready for startup, and cleanliness.

Readiness reviews of simple startups may involve only one person walking through the process to verify that nothing has changed and the equipment is ready to resume operation. Complex reviews may extend over many weeks or months as engineering, operations, and maintenance personnel verify equipment conformance to design intent, construction quality, procedure completion, training

PROCESS SAFETY BASICS 37

competency, and so forth. Typically, extensive checklists, multi-stage verification, and multiple functional sign-offs are required for startup authorization.

2.16 Conduct of Operations

Case Study. On January 21, 1997, an explosion and fire occurred at a refinery hydrocracker unit in Martinez, California, resulting in one death, 46 worker injuries, and an order for the surrounding community to shelter in place (Ref. 2.44). A temperature excursion occurred while attempting to recover from a process upset. Readings of a temperature data logger went unheeded because it had a history of unreliability. Radio transmissions from a field operator attempting to verify the temperatures locally were garbled. Thus, the control room operators did not depressurize the unit as required by procedure, and the overheated effluent piping burst. This incident illustrates how weaknesses in the conduct of operations can lead to tragedy. Continued operations in the face of unreliable or incomplete process information ultimately led to unreliable performance that exceeded safe operating limits.

Overview. *Conduct of operations* is the execution of operational and management tasks in a deliberate and structured manner. Conduct of Operations can be thought of as the real time manifestation and application of an organization's process safety culture. Conduct of operations includes, but is broader in scope than, just operating discipline. It involves first planning and documenting the work to be done and then executing according to the plan. ("Plan the Work--Work the Plan")

Conduct of operations is the day-to-day application of the process safety pillars assuring that:

- There are procedures for all aspect of operations
- The procedures are understood and followed all the time
- Equipment is maintained as needed
- Changes are controlled
- Audits and management reviews are effective and that findings are addressed.

In other words, an organization with a strong Conduct of Operations element has complete procedures and practices for all aspects of the process including operation and maintenance, and for controlling changes, that are followed all the time by all levels of the organization.

Operational Discipline is a subset of Conduct of Operations that deals with operational activities that produce consistent results. It is closely tied to an organization's culture. Conduct of operations institutionalizes the pursuit of excellence in the performance of every task and minimizes variations in performance. Personnel at every level are expected to perform their duties with alertness, due thought, full knowledge, sound judgment, and a sense of pride and accountability. Chapter 7 provides more detail on both an operator and engineer's role in Conduct of Operations.

A consistently high level of human performance is a critical aspect of any process safety program; indeed a less than adequate level of human performance will adversely impact all aspects of operations. As the complexity of operational activities increases, a commensurate increase in the formality of operations must also occur to ensure safe, reliable, and consistent performance of critical tasks.

2.17 Emergency Management

Case Study. On the morning of April 16, 1947, a fire was detected in a cargo hold aboard the freighter Grandcamp while it was moored near Texas City, Texas, loading ammonium nitrate fertilizer. At the time the fire started, the ship had been loaded with 1,400 tons of ammonium nitrate (in bags) in one hold and 800 tons of the same material in another hold. Over the next hour, the ship's captain decided to not use water to extinguish the fire, fearing that some of the cargo would be lost. Instead, the captain ordered sailors to close the hold and use high pressure steam to displace the oxygen. Unfortunately, once set afire, ammonium nitrate coated with paraffin does not need oxygen to burn. Pure ammonium nitrate will not burn, but when coated with a fuel such as a hydrocarbon wax, it can be heated by burning of the coating to its thermal decomposition temperature and an uncontrollable exothermic decomposition can be initiated. If the material is confined an explosion (deflagration and/or detonation can occur. The use of steam in a sealed ship hold in this situation further heated the solids and likely reduced the time to reach the point of "no return" from a runaway decomposition. The ship exploded, killing about 600 people, which, in terms of fatalities, is the worst industrial accident that has ever occurred in the U.S. The explosion aboard the Grandcamp also caused several other large explosions and fires over the next 16 hours involving a nearby chemical plant, nearby oil terminals, and ultimately led to a large explosion early in the morning of April 17 involving 1,000 tons of ammonium nitrate aboard the freighter Highflyer, which was moored in a slip adjacent to the Grandcamp (Figure 2.8).

According to one author who studied the Texas City disaster, "Safety and emergency preparedness . . . were grossly deficient, considering the enormity of the dangers. . . . Without preparations, little chance existed that anyone could cope with the effects of an accident quickly enough to prevent it from escalating into a disaster." Clearly, many failures occurred in the emergency response effort,

Figure 2.8. Aerial view of the burning Monsanto plant after the 1947 Texas City Disaster, (http://texashistory.unt.edu/ark:/67531/metapth11883) University of North Texas Libraries, The Portal to Texas History, crediting Moore Memorial Public Library, Texas City, Texas.

including the decision to use steam to displace oxygen rather than water to remove heat, and the failure to evacuate bystanders from the dock adjacent to the Grandcamp. In fact, hundreds of people gathered in the dock area between 8:00 a.m. and 9:12 a.m. that morning to get a better view of the brightly colored smoke and flames (Ref. 2.46).

Overview. *Emergency management* includes: 1) planning for possible emergencies, 2) providing resources to execute the plan, 3) practicing and continuously improving the plan, 4) training or informing employees, contractors, neighbors, and local authorities on what to do, how they will be notified, and how to report an emergency, and 5) effectively communicating with stakeholders in the event an incident does occur.

The scope of the *emergency* element extends well beyond "putting out the fire." This section focuses on three aspects of emergency planning and response:

- Protecting people, including people who are onsite, offsite, and emergency responders.
- Responding to catastrophic accidents involving explosions, large releases of chemicals, or other large releases of energy.
- Communicating with stakeholders, including neighbors and the media.

Emergency planning is typically performed by specialists, both within and external to the facility. Planners consult with the operations group and review work products from the *HIRA* element (section 2.8) to identify and select planning scenarios. Emergency response plans should be developed in concert with potentially involved or affected work groups, and they should be frequently reviewed with all potentially involved or affected workers. The operations group is typically responsible for immediate emergency response activities, such as shutting down the process and isolating hazardous material inventories, and they are assisted as quickly as possible by specially trained teams whose activities are coordinated by an incident commander. These teams often include facility sponsored response teams, outside agencies, including fire departments, medical responders, hazardous material (HAZMAT) teams, and, in some locations, mutual aid response teams from nearby facilities.

The following are the steps in developing an emergency response plan:

Identify accident scenarios based on hazards. Large-scale incidents involving chemicals generally present three classes of hazards: fires (thermal effects), explosions (pressure effects), and toxic vapor clouds (physiological effects). Often, the response plan can address a few scenarios involving each hazard class and cover the range of credible large-scale accident scenarios.

Assess credible accident scenarios. Evaluate the list of credible accident scenarios to determine the types and range of potential effects.

Select planning scenarios. Select planning scenarios based on the types of releases, the footprint or potentially affected area, and the incident history within the industry. Intended actions, such as evacuation versus shelter-in-place, should also be considered when selecting scenarios for inclusion in the emergency response plan.

Plan defensive response actions. Defensive response actions include, but are not limited to: emergency recognition and reporting, authority and methods for raising a standby or evacuation alarm, personal protective equipment to assist in evacuation, safe havens or shelter-in-place locations (and when this strategy should be used), evacuation routes, assembly points, and actions that should be taken at the assembly points, establishment of an emergency operations center

(EOC), including command and control of defensive actions and policies regarding transition of incident command authority.

Plan offensive response. Offensive response includes: firefighting preplans, defining the boundaries for the "hot" and "warm" zones, control of access into and out of these zones, response team communications, with particular emphasis on communications between responders, operations, and supporting groups, staffing and specific duties, guidance for PPE selection, decontamination procedures.

Develop written emergency response plan. Written emergency response plans spell out offensive response actions and form the basis for determining what facilities, equipment, staffing, training, communication, coordination, and other resources or activities are required.

Provide physical facilities and equipment. The resources needed for emergency response should be described in the response plan(s). Facilities should also give considerable thought to the location of equipment: siting it too far from likely response areas will increase response time, while siting it too close to locations where hazardous materials may be released can make it difficult impossible to use safely during an actual incident.

Maintain/test facilities and equipment. The response equipment must be maintained, periodically inventoried, and tested to ensure that it will function when it is needed.

Determine when unit operator response is appropriate. Plans should include when it is appropriate for operators to address the event and what they should and should not do.

Train the Emergency Response Team (ERT) members. The ERT not only needs initial training, but refresher training and drills on a regular basis.

Plan communications. This is normally a very simple task for facility employees, and becomes progressively more difficult with contractors, local authorities, neighbors, and other stakeholders.

Inform and train all personnel. Individuals who might be affected by an emergency should be trained or notified on how they will be alerted of an emergency, what actions they may be asked to take, and what to do to protect themselves.

Periodically review emergency response plan. Emergency response plans are particularly prone to becoming out of date, because unlike operating procedures, workers are less likely to get them out and read them on a routine basis.

Pillar: Learn from Experience

There are four elements to the pillar of Commit to Process Safety

- Incident Investigation (Section 2.18)
- Measurement and Metrics (Section 2.19)
- Auditing (Section 2.20)
- Management Review and Continuous Improvement (Section 2.21)

2.18 Incident Investigation

Case Study. The incident below comes from outside the process industries, but is again illustrative of how these management principles can apply to other endeavors.

The in-flight failure of the Columbia Space Shuttle on January 16, 2003, which resulted in the deaths of the seven-member crew, is a classic example of a failure to adequately investigate and address abnormal system performance. To protect the shuttle's fragile thermal protection system, the shuttle design specifications required no shedding of foam insulation. Despite this requirement, at least 65 missions had experienced foam loss, including six instances of foam shedding from the bipod ramp, the same location that ultimately caused the loss of the Columbia. For example, on October 7, 2002, during launch of STS-112, foam loss caused 707 dings in the thermal protection system, 298 of which were greater than an inch in one dimension. One piece of debris knocked off a thermal protection tile, exposing the orbiter's skin to the heat of re-entry. Fortunately, the missing tile on STS-112 happened to be at the location of a thick aluminum plate, so a burn-through did not occur. Despite these near miss incidents and extensive experience that was contrary to the design requirements of the shuttle, NASA did not thoroughly investigate the root causes of foam separation. Experience with successful missions had lulled NASA into believing that the shuttle was immune from damage from shedding debris. Proper investigation of these near miss incidents could have prompted NASA to change the shuttle design to prevent the shedding of the foam, to decrease the shuttle's susceptibility to a foam strike, or both (Refs. 2.51 and 2.52). The lessons learned from the previous Challenger tragedy in 1986 concerning "normalization of deviance" with the "O" ring seals had to be relearned again in 2003, in this case for deviation from design intent and

Emergency Planning and Response are part of the OSHA PSM and EPA RMP regulations (see Table 2.1) and are covered by local regulations and other countries.

performance specifications for the foam. The Columbia incident is described in more detail in Section 3.3.

Overview. *Incident investigation* is a process for reporting, tracking, and investigating incidents and near misses that includes a formal process for conducting incident investigations, including staffing, performing, documenting, and tracking investigations of process safety incidents. It also includes the trending of incident and incident investigation data to identify recurring incidents. The purpose of incident investigation is to identify and eliminate the causes of incidents to improve organizational performance. The *incident investigation* process also manages the resolution and documentation of recommendations generated by the investigations.

Incident investigation is a way of learning from incidents that occur over the life of a facility or enterprise and communicating the lessons learned to both internal personnel and other stakeholders. Depending upon the depth of the analysis, this feedback can apply to the specific incident under investigation or a group of incidents sharing similar root causes at one or more facilities.

One error in the investigation process is to use the investigation as a means to assign blame to personnel involved in an incident. Incident investigation should not be used to assign blame. Assigning blame results in failure to get at the management system failures that caused the incident and in ineffective recommendations being implemented. A more effective approach is to develop recommendations that address the management causes of the incidents.

Many companies are adopting the CCPS *Process Safety Leading and Lagging Metrics* (Ref. 2.47) or API RP 754 *Process Safety Performance Indicators for the Refining & Petrochemical Industries* (Ref. 2.48) to define incidents, events, near misses and unsafe behaviors. Figure 2.9 shows the Process Safety Metric Pyramid taken from the CCPS document. At the top are Process Safety Incidents. Incidents meet a threshold criteria set forth in the API 754 document, for example, a release of more than 50 kg (110 lb.) of a liquefied flammable gas or liquid with a boiling point ≤ 35 °C (95 °F) and a flash point ≤ 23 °C (73 °F), or an incident that resulted in a fatality.

The next level in the pyramid is a Process Safety Event, which does not meet the reporting threshold of an incident, but are precursors to future incidents.

The Near Miss represents minor Loss of Primary Containment events or other system failures. Examples are challenges to a safety system (even though the system worked), alarms or interlocks disabled without proper authorization, unusual line-ups on shift, bypassed or out-of-service equipment, or an equipment inspection with a result outside of acceptable limits (e.g. pipe wall thickness too low).

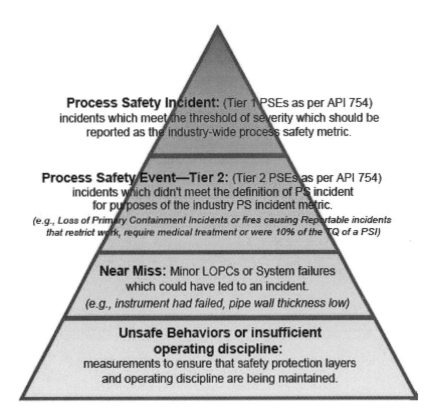

Figure 2.9. CCPS and API Process Safety Metric Pyramid (Ref. 2.46).

A company may opt for a different set of definitions, but, in any case, the engineer's role in terms of training, monitoring and follow-up is the same.

Once an incident occurs, there are several types of incident investigation methods, described in *Guidelines for Investigating Chemical Process Incidents, Second Edition* (Ref 2.49), that can be used. The method used will usually depend on the perceived severity of the incident. These methods can range from simple brainstorming to use of checklists to creating logic trees. The investigation team should consist of people with expertise in the investigation method and with process expertise appropriate to the event. For example, the investigation into the pipe rupture described in Section 2.11 should have a metallurgist and someone familiar with the company's mechanical integrity program. Once completed, recommendations need to be implemented and lessons learned should be shared with similar facilities within the organization. Finally, companies should also track

PROCESS SAFETY BASICS   45

incidents and near misses in a database to enable them to analyze the events for trends that can be causing repeat incidents.

A new engineer in a company is very likely to be called into an investigation, especially if working in a production facility. Incident investigation is part of the OSHA PSM and EPA RMP regulations (see Table 2.1) and is covered by local regulations and other countries.

2.19 Measurement and Metrics

Case Study. On September 23, 1999, the Mars Climate Orbiter was lost as it attempted to enter orbit (Ref. 2.49). During its 9-month journey, propulsion maneuvers were required ten times more often than were expected by the navigation team. After the spacecraft loss, investigators discovered that the trajectory errors were introduced by a software module that had been coded in the wrong measurement units. Had the unexpected deviations in trajectory been investigated during the flight, the loss of the orbiter might have been avoided. Like flight trajectories, management system metrics provide data about actual system performance versus intended system performance. If the underlying causes of deviations are not investigated, understood, and corrected, the "flight" of the management system may continue to appear normal for months or years until a critical maneuver, such as a unit startup or retirement of a key worker, results in a catastrophic system failure.

Overview. The *metrics* element establishes performance and efficacy indicators to monitor the effectiveness of the RBPS management system and its constituent elements and work activities. This element addresses which indicators to consider, how often to collect data, and what to do with the information to help ensure responsive, effective RBPS management system operation.

Fortunately, serious process safety accidents occur relatively infrequently. However, when they do occur, they usually involve a confluence of root causes, some of which involve degraded effectiveness of management systems or, worse, complete failure of management system activities. Facilities should monitor the real-time performance of management system activities rather than wait for accidents to happen or for infrequent audits to identify latent management system failures. Such performance monitoring allows problems to be identified and corrective actions to be taken before a serious incident occurs.

A combination of leading and lagging indicators is often the best way to provide a complete picture of process safety effectiveness. A leading indicator measures the performance of process safety management systems whose failure that can contribute to incidents before an incident occurs. Examples are the performance of:

- The Asset Integrity and Reliability system, e.g., are inspections taking place on time, are deficiencies corrected?
- The HIRA system, e.g. are hazard studies done on time, are the recommendations implemented in a timely manner?
- The Training and Performance system, e.g., has everyone received their refresher training on time?

Lagging indicators are the incidents and near misses themselves. An example of a lagging indicator is an incident rate. Lagging indicators are generally not sensitive enough to be useful for continuous improvement of process safety management systems because incidents occur too infrequently. The CCPS has developed a list of suggested leading and lagging indicators (Ref 2.48).

Measuring process safety management performance requires the use of leading indicators. Some examples of leading indicators are:

- Percent of HIRA updates that are done on time.
- Percent of recommendations from HIRA or MOC reviews that are completed within a specified time.
- Percentage of recommendations from HIRA or MOC reviews that are past due.
- Percentage of MOCs executed properly.
- Percentage of Safe Work Permits executed properly.
- Percent of inspections that are done on time, or conversely, past due.

As a new engineer, you may be involved in collecting and even analyzing the data for leading and lagging indicators.

2.20 Auditing

Case Study. On September 25, 1998, an explosion occurred at the Longford gas plant (see Section 2.8). An audit conducted by a corporate team six months prior to the explosion had determined that the gas plant was successfully implementing its process safety management system. However, a Royal Commission subsequently investigated the explosion and found significant deficiencies in the areas of risk identification, analysis, and management, training, operating procedures, documentation, and communications. These long-standing problems had not been detected by a series of prior audits. (Figure 2.10)

Overview. An *audit* is a systematic, independent review to verify conformance with prescribed standards of care. The purpose of auditing is to identify

PROCESS SAFETY BASICS

Figure 2.10. Photograph of failed end of heat exchanger, Ref. 2.33.

management system and performance gaps in the process safety management system and correct those gaps before an incident occurs. *Auditing* employs a well-defined review process to ensure consistency and to allow the auditor to reach defensible conclusions. It complements other RBPS control and monitoring activities in elements such as metrics (Section 19), management review (Section 21), and inspection work activities that are part of the asset integrity and conduct of operations elements (Sections 2.11 and 16).

An audit involves a methodical, typically team-based, assessment of the implementation status of one or more RBPS elements against established requirements, normally directed by the use of a written protocol. Data are gathered through the review of program documentation and implementation records, direct observations of conditions and activities, and interviews with individuals having responsibilities for implementation or oversight of the element(s) or who might be affected by the RBPS management system. The data are analyzed to assess compliance with requirements, and the conclusions and recommendations are documented in a written report.

Audits can be conducted by a team of qualified personnel selected from a variety of sources, depending upon the scope, needs, and other aspects of the specific situation. Team members might be selected from staff at the facility being audited, from other company locations (e.g., from another operating facility or from corporate staff functions), or from outside the company, for example, a consulting firm. As a new engineer it is unlikely that you will actually be conducting an audit, but you may be involved in interviews and responsible for providing information and documentation to an auditor.

Auditing is one of the elements in the OSHA PSM and EPA RMP regulations, and audits are audits of covered processes are required at least every three years. Auditing, and the frequency of audits, can be dictated by other local and other national codes.

2.21 Management Review and Continuous Improvement

Case Study. The accident at the Motiva refinery described in Section 2.10 resulted from a confluence of several factors, as is commonly the case in large industry accidents:

- The corrosivity and flammability hazards associated with changing the tank from fresh acid service to spent acid service were not identified and controlled
- Repeated requests for tank inspections and repairs were deferred or ignored
- The hot work permit failed to specify atmospheric monitoring despite previous permit denials because of toxic and flammable gas concentrations.

Each of these contributing factors was the result of a management system breakdown (management of change, asset integrity, and safe work) that could have been identified and corrected by timely management review.

Overview. *Management review* is the routine evaluation of whether management systems are performing as intended and producing the desired results as efficiently as possible. Management reviews have many of the characteristics of an audit as described in Section 2.20. They require a similar system for scheduling, staffing, and effectively evaluating all RBPS elements, and a system should be in place for implementing any resulting plans for improvement or corrective action and verifying their effectiveness. However, because the objective of a management review is to spot current or incipient deficiencies, the reviews are more broadly focused and more frequent than audits, and they are typically conducted in a less formal manner.

Effective performance is a critical aspect of any process safety program; however, a breakdown or inefficiency in a safety management system may not be immediately obvious. For example, if a facility's training coordinator unexpectedly departed, required training activities might be disrupted. The existing trained workers would undoubtedly continue to operate the process, so there would be no outward appearance of a deficiency. An audit or incident might eventually reveal any incomplete or overdue training, but by then it could be too late. The *management review* process provides regular checkups on the health of process safety management systems in order to identify and correct any current or incipient deficiencies before they might be revealed by an audit or incident.

2.22 Summary

This chapter has introduced the concept of management systems, risk, and risk based process safety. RBPS essentially states that the degree of rigor used in implementing the management system elements is a function of the risk of the process. The elements of a RBPS program, as enumerated by the CCPS, have been introduced and compared to the existing OSHA and EPA process safety management regulations. Your role with respect to some of these elements, as a new engineer in an organization, has been briefly described.

These elements work together like the interlocking pieces of a jigsaw puzzle. Let's use a hypothetical situation to illustrate this.

You are an engineer in a chemical facility when an incident or serious near miss has occurred. An *incident investigation* must occur. Near misses are much more likely to be noticed and reported in a company with a good *process safety culture*. A good investigation will use appropriate resources in the facility and other expertise within or outside of the company as needed (*workforce involvement*). Recommendations will come from the investigation and the results will be shared with similar facilities in the company so everyone can learn from the event.

Albert Einstein is credited with defining insanity as doing the same thing over and over again and expecting different results. Therefore, if you do not want this event to occur again, it follows that the recommendations from the investigation will involve changes in what the facility is doing. These can range from procedural changes to the addition of new process safety controls or interlocks to a redesign of the process, including new process equipment. In any case, you are going to change something (*management of change*). An MOC review needs to be held, again with people with the appropriate expertise. If it is an extensive change, you may decide to update the HIRA (*Hazard Identification and Risk Analysis*), especially if an update is due in a relatively short time, or if the change might be expected to introduce brand new hazards that the current safeguards are not designed for. During or even before the MOC review, you need to make sure the resulting process will comply with standards (*compliance with standards*). New or updated procedures need to be written (*operating procedures*). If new equipment, even as simple as a new safety interlock, is added it needs to be put on a maintenance and inspection schedule (*asset integrity*). Operators and maybe maintenance personnel need to be trained on the new procedure, process, etc. (*training and performance assurance*). Changes to piping and equipment need to be done in a safe manner (*safe work practices*). In addition to information about the event itself, new information, such as piping and instrumentation diagrams, equipment or technology descriptions need to be added to the existing process safety information documentation (*process knowledge management*). Before you restart the process with the changes implemented a pre-startup safety review has to be held (*operational readiness*). When the entire project is complete, the company may choose to do a review of the project to be sure all of its procedures and processes were complied with, and if not, why (*measurement and metrics*).

This hypothetical example did not cover every element, but it did cover more than half of them, illustrating how all the pieces come together.

2.23 References

2.1 Process Safety Management (Control of Acute Hazards), Chemical Manufacturers Association, Washington, DC., 1985

2.2 Guidelines for Technical Management of Chemical Process Safety, Center for Chemical Process Safety, New York, 1989.

2.3 Management of Process Hazards, API Recommended Practice 750 (Not Active), American Petroleum Institute, Washington D.C. 1990

2.4 Guidelines for Risk Based Process Safety, Center for Chemical Process Safety, New York, 2007.

2.5 Combustible Dust Fire and Explosions at CTA Acoustics, Inc. Corbin, Kentucky February 20, 2003; U.S. Chemical Safety and Hazard Investigation Board, February 15, 2005. http://www.csb.gov/index.cfm?folder=completed_investigations&page=info&INV_ID=35

2.6 American Petroleum Institute, 1220 L Street, NW, Washington, DC 20005. www.api.org

2.7 American Chemistry Council 1300 Wilson Blvd., Arlington, VA 22209. www.americanchemistry.com

2.8 ISO 12001 – Environmental Management System, International Organization for Standardization (ISO), Geneva, Switzerland. www.iso.org/iso/en/iso9000-12000/index.html

2.9 OHSAS 18001 – International Occupational Health and Safety Management System. www.ohsas-18001-occupational-health-and-safety.com/

2.10 Organization for Economic Cooperation and Development – Guiding Principles on Chemical Accident Prevention, Preparedness, and Response, 2nd edition, 2003, Organization for Economic Co-Operation and Development, Paris, 2003. (www2.oecd.org/ guidingprinciples/index.asp)

2.11 American National Standards Institute, 25 West 23rd Street, New York, NY 10036. (www.ansi.org)

2.12 American Society of Mechanical Engineers, Three Park Avenue, New York, NY 10016. www.asme.org

2.13 The Chlorine Institute, 1300 Wilson Blvd., Arlington, VA 22209, www.chlorineinstitute.org

2.14 The Instrumentation, Systems, and Automation Society, 67 Alexander Drive, Research Triangle Park, NC 27709. www.isa.org

2.15 National Fire Protection Association, 1 Batterymarch Park, Quincy, MA, 02169. www.nfpa.org

2.16 Process Safety Management of Highly Hazardous Chemicals (29 CFR 1910.119), U.S. Occupational Safety and Health Administration, May 1992. www.osha.gov

2.17 Section 5(a)(1) – General Duty Clause, Occupational Safety and Health Act of 1970, Public Law 91-596, 29 USC 652, December 29, 1970. www.osha.gov

2.18 Accidental Release Prevention Requirements: Risk Management Programs Under Clean Air Act Section 112(r)(7), 20 CFR 68, U.S. Environmental Protection Agency, June 20, 1996 Fed. Reg. Vol. 61[31667-31730]. www.epa.gov

2.19 Clean Air Act Section 112(r)(1) – Prevention of Accidental Releases – Purpose and general duty, Public Law No. 101-529, November 1990. www.epa.gov

2.20 California Accidental Release Program (CalARP) Regulation, CCR Title 19, Division 2 – Office of Emergency Services, Chapter 2.5, June 28, 2002. www.oes.ca.gov

2.21 Toxic Catastrophe Prevention Act (TCPA), New Jersey Department of Environmental Protection Bureau of Chemical Release Information and Prevention, N.J.A.C. 7:31 Consolidated Rule Document, April 17, 2006. www.nj.gov/dep

2.22 Contra Costa County Industrial Safety Ordinance. www.co.contra-costa.ca.us/

2.23 Extremely Hazardous Substances Risk Management Act, Regulation 1201, Accidental Release Prevention Regulation, Delaware Department of Natural Resources and Environmental Control, March 11, 2006. www.dnrec.delaware.gov/

2.24 Chemical Accident Prevention Program (CAPP), Nevada Division of Environmental Protection, NRS 259.380, February 15, 2005. http://ndep.nv.gov/bapc/capp/

2.25 Australian National Standard for the Control of Major Hazard Facilities, NOHSC: 1012, 2002. www.docep.wa.gov.au/

2.26 Environmental Emergency Regulations (SOR/2003-307), Environment Canada. www.ec.gc.ca/CEPARegistry/regulations/detailReg.cfm?intReg=70

2.27 Control of Major-Accident Hazards Involving Dangerous Substances, European Directive Seveso II (96/82/EC). http://europa.eu.int/comm/environment/seveso/

2.28 Korean OSHA PSM standard, Industrial Safety and Health Act – Article 20, Preparation of Safety and Health Management Regulations. Korean Ministry of Environment – Framework Plan on Hazardous Chemicals Management, 2001-2005.
www.kosha.net/jsp/board/viewlist.jsp?cf=29099&x=19565&no=3

2.29 Malaysia – Department of Occupational Safety and Health (DOSH) Ministry of Human Resources Malaysia, Section 16 of Act 512. http://dosh.mohr.gov.my/

2.30 Mexican Integral Security and Environmental Management System (SIASPA), 1998. www.pepsonline.org/Publications/pemex.pdf

2.31 Control of Major Accident Hazards Regulations (COMAH), United Kingdom

2.32 EPA/OSHA Joint Chemical Accident Investigation Report, Napp Technologies, Inc., Lodi, New Jersey, EPA 550-R-97-002, United States Environmental Protection Agency, October 1997. http://www.epa.gov/oem/docs/chem/napp.pdf

2.33 Report of the Longford Royal Commission, Government Printer for the State of Victoria, 1999. (http://www.parliament.vic.gov.au/papers/govpub/VPARL1998-99No61.pdf)

2.34 Guidelines for Hazard Evaluation Procedures (Third Edition with Worked Examples), Center for Chemical Process Safety, New York, 2008.

2.35 Layer of Protection Analysis – Simplified Process Risk Analysis, Center for Chemical Process Safety, New York, 2001.

2.36 Guidelines for Chemical Process Quantitative Risk Analysis (Second Edition), Center for Chemical Process Safety, New York, 1999.

2.37 Guidelines for Initiating Events and Independent Protection Layers, Center for Chemical Process Safety, New York, 2015.

2.38 Chemical Accident Investigation Report – Terra Industries, Inc. Nitrogen Fertilizer Facility, Port Neal, Iowa, U.S. Environmental Protection Agency, Region 7, Emergency Response and Removal Branch, Kansas City, Kansas, issued January, 1996.

2.39 U.S. Chemical Safety and Hazard Investigation Board, Case Study, Report No. 2001-05-I-DE, Refinery Incident, Motive Enterprises LLC Delaware City Refinery, Delaware City, DE, July 17, 2001. (http://www.csb.gov/motiva-enterprises-sulfuric-acid-tank-explosion/)

2.40 U.S. Chemical Safety and Hazard Investigation Board, Interim Investigation Report, Chevron Richmond Refinery Fire, Chevron Richmond Refinery, Richmond, CA, August 6, 2012. (http://www.csb.gov/assets/1/19/Chevron_Interim_Report_Final_2013-04-17.pdf)

2.41 Incidents That Define Process Safety, Center for Chemical Process Safety, New York, 2008.

2.42 M. Elisabeth Pate-Cornell, Learning from the Piper Alpha Accident: A Postmortem Analysis of Technical and Organizational Factors, Risk Analysis, Vol. 13, No. 2, 1993, p. 215-232.

2.43 Lees, Frank P., Loss Prevention in the Process Industries, Hazard Identification, Assessment, and Control, 2nd edition, Butterworth-Heinemann, Oxford, England, 1996.

2.44 Guidelines for Managing Process Safety during Organizational Change, Center for Chemical Process Safety, New York, 1999.

2.45 EPA Chemical Accident Investigation Report – Tosco Avon Refinery, Martinez, California, EPA550-R-98-009, U.S. Environmental Protection Agency Chemical Emergency Preparedness and Prevention Office, Washington, DC, 1998.

2.46 Stephens, Hugh W., The Texas City Disaster, 1947, University of Texas Press, Austin, TX, 1997.

2.47 Center for Chemical Process Safety (CCPS), Process Safety Leading and Lagging Metrics – You Don't Improve What You Don't Measure, January 2011, (http://www.aiche.org/sites/default/files/docs/pages/metrics%20english%20updated.pdf)

2.48 American Petroleum Institute, ANSI/API Recommended Practice 754, Process Safety Performance Indicators for the Refining and Petrochemical Industries, First Edition, Washington D.C., 2010.

2.49 Mars Climate Orbiter Mishap Investigation Board – Phase I Report, National Aeronautics and Space Administration, Washington D.C. November 10, 1999

3

The Need for Process Safety

Failures of process safety management systems are deadly and costly. Over time major process safety incidents bring significant public attention to the process industries and to the need for process safety. For example, the founding of CCPS, the publisher of this book (and about 120 other books on process safety), was an industry response to the Methyl Isocyanate release at Bhopal, India in 1984 (Section 3.15) which resulted in the death of over 3,000 people and injured tens of thousands, with some estimates even higher. A fire and explosion at a PEMEX LPG terminal in Mexico City (Section 3.14), also in 1984, killed over 600 people and injured about 7,000. Major environmental damage has also been caused by process safety incidents. The firefighting efforts during a fire in a Sandoz warehouse in Basel, Switzerland in 1986 (Section 3.22) caused the release of many different chemicals, including many pesticides, which resulted in massive destruction to aquatic life in the Rhine for a distance of up to 400 km (250 miles). Fishing was banned for 6 months.

Each event also increases public awareness to the potential dangers of chemical and petrochemical plants resulting in less tolerance for such events. Just consider the around the clock news and publicity that happened after the Exxon Valdez grounding and oil spill (Section 3.13) or the BP Macondo Well fire and oil spill in the Gulf of Mexico (Section 3.16). In addition to the negative publicity and damage to each company's image, cleanup costs and fines ran into the billions of dollars for each of these incidents. The lessons learned from these events have also helped develop and expand process safety concepts to the point they are currently at today and continue to drive growth and understanding of process safety management concepts.

Students and new engineers can find reports of incidents from several sources. An extremely useful source is the Chemical Safety Board (CSB) (www.csb.gov). The CSB is a US government agency charged with investigating chemical accidents at industrial facilities. The reports of their investigations are available for download from the CSB website. Also, the CSB has created a series of videos that describe many process safety incidents. As of 2015, there are over 70 reports and 30 videos available (videos are available on both the CSB website and YouTube). See Appendix B for a list of CSB videos.

The CCPS book, *Incidents That Define Process Safety* (IDPS) (Ref. 3.1), describes many more events. The IDPS book also describes events from other industries besides chemical and petrochemical, illustrating the point that many process safety elements are universal in their relevance to safe operations.

Another good source of incident descriptions is *Lees' Loss Prevention in the Process Industries* (Ref. 3.2). This three volume reference lists many accidents in its appendices. The descriptions are more technical than the CSB publications and might also be useful for students and new engineers. Lee's may be found in your Chemical Engineering department's library.

One more source of incident descriptions and learnings is the CCPS Process Safety Beacon. The Process Safety Beacon is aimed at delivering process safety messages to plant operators and other manufacturing personnel. The Beacon is issued monthly. Each issue presents a real-life incident and describes the lessons learned and practical means to prevent similar incidents in your plant. Students can gain access to past issues of the Beacon though the Safety and Chemical Engineering Education (SAChE) archives online at (http://sache.org/beacon/products.asp)

The next sections describe some selected events extracted from the IDPS book, CSB investigations and other sources. After a description of the event, a few Risk Based Process Safety Elements (Chapter 2), whose failure is involved in the incident, are identified in a Key Lessons section. Table 3.1 lists the RBPS management system highlighted in each incident.

Table 3.1 Selected incidents and Process Safety Management systems.

	Pillars of Risk Based Process Safety			
Incident	Commit to Process Safety	Understand Hazards and Risk	Managing Risk	Learning from Experience
BP Explosion, Texas City, 2005	Process Safety Culture	Process Knowledge Management Training & Performance Assurance	Management of Change Asset Integrity & Reliability	
Arco Channelview Explosion, 1990			Management of Change Asset Integrity & Reliability Conduct of Operations	

Table 3.1 Selected incidents and Process Safety Management systems, continued.

Incident	Pillars of Risk Based Process Safety			
	Commit to Process Safety	Understand Hazards and Risk	Managing Risk	Learning from Experience
Space Shuttle Columbia, 2003	Process Safety Culture	Hazard Identification & Risk Management		
Concept Sciences Explosion, 1999		Process Knowledge Management Hazard Identification & Risk Management		
Esso Longford Gas Plant Explosion, 1998	Process Safety Competency	Hazard Identification & Risk Management	Management of Change	
Port Neal, Ammonium Nitrate Explosion, 1994		Hazard Identification & Risk Management	Operating Procedures	
Piper Alpha, 1988			Safe Work Practices Emergency Management	
Partridge Raleigh Oilfield Explosion, 2006	Compliance with Standards		Safe Work Practices Contractor Management	
Texaco, Milford Haven Explosion, 1994			Asset Integrity and Reliability	

Table 3.1 Selected incidents and Process Safety Management systems, continued

	Pillars of Risk Based Process Safety			
Incident	Commit to Process Safety	Understand Hazards and Risk	Managing Risk	Learning from Experience
Formosa Plastics VCM Explosion, 2004			Conduct of Operations Management of Organizational Change Training and Performance Emergency Management	
Flixborough Explosion, 1974	Compliance with Standards		Management of Change	
Sandoz Warehouse Fire, 1986			Emergency Management	
Exxon Valdez, 1989			Conduct of Operations Management of Change	
PEMEX LPG Explosion, 1984	Compliance with Standards			
Macondo Well, 2010	Process Safety Culture			Incident Investigation
Bhopal Methyl Isocyanate Release, 1984	Process Safety Culture	Hazard Identification & Risk Assessment	Management of Change	

The Swiss Cheese model. In the descriptions you will generally find that all the incidents have several failures leading to the final outcome. This phenomenon is frequently described by what is known as the "Swiss Cheese" model. For many years, safety experts have used the Swiss Cheese model proposed by James Reason, (Ref. 3.3.) as one theory to explain catastrophic incidents.

NEED FOR PROCESS SAFETY

The model can help managers and workers in process industries understand the events, failures, and decisions that can cause an incident or near miss to occur. The example in Figure 3.1 depicts layers of protection that all have holes. When a set of unique circumstances occur, the holes line up and allow an incident to happen. We display protection layers in the figure as slices of cheese. The holes in the cheese represent the following potential failures in the protection layers:

- Human errors during design, construction, commissioning, operation or maintenance
- Management decisions
- Single-point equipment failures or malfunctions
- Knowledge deficiencies
- Management system inadequacies, such as a failure to perform hazard analyses, failure to recognize and manage changes, or inadequate follow-up on previously experienced incident warning signs.

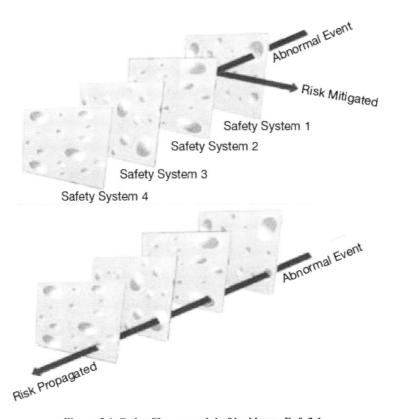

Figure 3.1. Swiss Cheese model of incidents, Ref. 3.1.

As Figure 3.1 illustrates, incidents are typically the result of multiple failures to address hazards effectively. A management system may include physical safety devices or planned activities that protect and guard against failure. An effective process safety management system can reduce the number of holes and the size of the holes in each of the systems layers.

Normally references are placed at the end of the chapter. In this chapter, relevant references are listed after each incident. This is done to allow you to research the incident further if you want to know more.

3.1 Process Safety Culture: BP Refinery Explosion, Texas City, 2005

3.1.1 Summary

An explosion occurred within the Isomerization Unit (ISOM) of BP's Texas City Refinery during a startup after a turnaround in March 2005. Fifteen contractors were killed and over 170 people harmed. There was major damage to the ISOM and adjacent plant and equipment.

The portable plant buildings where the contractors were located were being used to support an adjacent plant turnaround. They were in an area operated as an uncontrolled area, i.e., a safe area without any Hot Work Permit/Electrical controls imposed (CCPS, 2008).

3.1.2 Detailed Description

See Figure 3.2 for a diagram of the ISOM system. The following excerpt from the CSB report describes the incident:

> During the startup, operations personnel pumped flammable liquid hydrocarbons into a distillation tower for over three hours without any liquid being removed, which was contrary to startup procedure instructions. Critical alarms and control instrumentation provided false indications that failed to alert the operators of the high level in the tower. Consequently, unknown to the operations crew, the 170-foot (52-m) tall tower was overfilled and liquid overflowed into the overhead pipe at the top of the tower.
>
> The overhead pipe ran down the side of the tower to pressure relief valves located 148 feet (45 m) below. As the pipe filled with liquid, the pressure at the bottom rose rapidly from about 21 pounds per square inch (psi) to about 64 psi. The three pressure relief valves opened for six minutes, discharging a large quantity of flammable liquid to a blowdown drum

NEED FOR PROCESS SAFETY

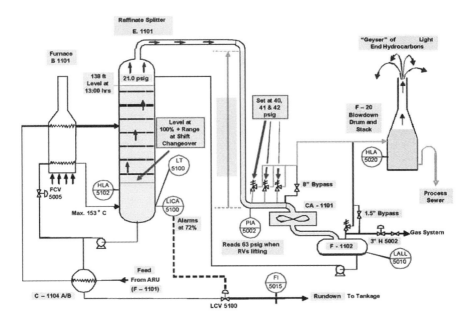

Figure 3.2. Process flow diagram of the Raffinate Column and blowdown drum, source (CCPS, 2008).

with a vent stack open to the atmosphere. The blowdown drum and stack overfilled with flammable liquid, which led to a geyser-like release out the 113-foot (34 m) tall stack. This blowdown system was an antiquated and unsafe design; it was originally installed in the 1950s, and had never been connected to a flare system to safely contain liquids and combust flammable vapors released from the process.

The released volatile liquid evaporated as it fell to the ground and formed a flammable vapor cloud. The most likely source of ignition for the vapor cloud was backfire from an idling diesel pickup truck located about 25 feet (7.6 m) from the blowdown drum. The 15 employees killed in the explosion were contractors working in and around temporary trailers that had been previously sited by BP as close as 121 feet (37 m) from the blowdown drum (CSB, 2007).

Figures 3.3 and 3.4 show the damage to the unit and the portable buildings, respectively.

3.1.3 Causes

The BP investigation concluded "that while many departures to the startup procedure occurred, the key step that was instrumental in leading to the incident

Figure 3.3. Texas City Isom Unit aftermath, courtesy CSB.

Figure 3.4. Portable buildings destroyed where contractors were located, courtesy CSB.

was the failure to establish Heavy Raffinate rundown to tankage, while continuing to feed and heat the tower. By the time the Heavy Raffinate flow was eventually started, the Splitter bottoms temperature was so high, and the liquid level in the tower so high, that this intervention made matters worse by introducing significant additional heat to the feed." (CCPS, 2008).

The investigation team concluded that the Splitter was overfilled and overheated because "the Shift Board Operator did not adequately understand the process or the potential consequences of his actions or inactions on March 23."

3.1.4 Key Lessons

Many Risk Based Process Safety Elements were involved in the BP Texas City explosion. Five are listed here. The bulleted findings below are taken directly from the CSB report unless noted otherwise.

Process Safety Culture (Section 2.2). Process safety culture is the first of 20 risk based process safety elements (see Chapter 2). Perhaps most striking of the CSB findings are those with respect to the process safety culture at BP and the Texas City plant. Listed below are some of the CSB's findings regarding BP's process safety culture. Some of these findings could easily apply to other companies. The CSB recommended that BP create an "independent panel of experts to examine BP's corporate safety management systems, safety culture, and oversight of the North American refineries." This became known as the Baker Panel. The Baker Panel report focused on safety management systems at BP and resulted in ten recommendations to the BP Board of Directors (BP Review Panel, 2007).

Selected CSB findings:

- "Cost-cutting, failure to invest and production pressures from BP Group executive managers impaired process safety performance at Texas City.
- The BP Board of Directors did not provide effective oversight of BP's safety culture and major incident prevention programs. The Board did not have a member responsible for assessing and verifying the performance of BP's major incident hazard prevention programs.
- Reliance on the low personal injury rate at Texas City as a safety indicator failed to provide a true picture of process safety performance and the health of the safety culture.
- A "check the box" mentality was prevalent at Texas City, where personnel completed paperwork and checked off on safety policy and procedural requirements even when those requirements had not been met."

Selected Baker Panel finding:

- "BP has not instilled a common, unifying process safety culture among its U.S. refineries. Each refinery has its own separate and distinct process safety culture. While some refineries are far more effective than others in promoting process safety, significant process safety culture issues exist at all five U.S. refineries, not just Texas City. Although the five refineries do not share a unified process safety culture, each exhibits some similar weaknesses. The Panel found instances of a lack of operating discipline, toleration of serious deviations from safe operating practices, and apparent complacency toward serious process safety risks at each refinery."

Process Knowledge Management (Section 2.7). BP acquired the Texas City refinery as part of its merger with Amoco in 1999. Neither Amoco (the previous facility operator) nor BP replaced blowdown drums and atmospheric stacks, even though a series of incidents warned that this equipment was unsafe. In 1992, OSHA cited a similar blowdown drum and stack as unsafe, but the citation was withdrawn as part of a settlement agreement and therefore the drum was not connected to a flare as recommended. Amoco, and later BP, had safety standards requiring that blowdown stacks be replaced with equipment such as a flare when major modifications were made. In 1997, a major modification replaced the ISOM blowdown drum and stack with similar equipment but Amoco did not connect it to a flare. In 2002, BP engineers proposed connecting the ISOM blowdown system to a flare, but a less expensive option was chosen.

Training and Performance Assurance (Section 2.13)

- A lack of supervisory oversight and technically trained personnel during the startup, an especially hazardous period, was an omission contrary to BP safety guidelines. An extra board operator was not assigned to assist, despite a staffing assessment that recommended an additional board operator for all ISOM startups.
- Supervisors and operators poorly communicated critical information regarding the startup during the shift turnover; BP did not have a shift turnover communication requirement for its operations staff. ISOM operators were likely fatigued from working 12-hour shifts for 29 or more consecutive days.
- The operator training program was inadequate. The central training department staff had been reduced from 28 to eight, and simulators were unavailable for operators to practice handling abnormal situations, including infrequent and high hazard operations such as startups and unit upsets.

Management of Change (Section 2.14)

- BP Texas City did not effectively assess changes involving people, policies, or the organization that could impact process safety. For example, the control room staff was reduced from 2 people to one, who was overseeing three units.
- Local site Management of Change rules required that where a portable building was to be placed within 100 meters (350 ft) of a process unit a Facility Siting Analysis had to be carried out. However, this location had already been used many times for these trailers. Not doing an effective MOC put all the people in the portable buildings at unnecessary risk (CCPS, 2008).

Asset Integrity and Reliability (Section 2.11)

- The process unit was started despite previously reported malfunctions of the tower level indicator, level sight glass, and a pressure control valve.
- Deficiencies in BP's mechanical integrity program resulted in the "run to failure" of process equipment at Texas City.

3.1.5 References and Links to Investigation Reports

- CSB, 2007. U.S. Chemical Safety and Hazard Investigation Board, Investigation Report, Report No. 2005-04-I-TX, Refinery Explosion and Fire. BP Texas City, Texas. March 23, 2007(http://www.csb.gov/investigations).
- U.S. Chemical Safety and Hazard Investigation Board, Video - Anatomy of a Disaster, (http://www.csb.gov/videos).
- BP Review Panel, 2007. The Report of the BP U.S. Refineries Independent Safety Review Panel, January 2007, (Baker Panel). (http://www.bp.com/liveassets/bp_internet/globalbp/globalbp_uk_english/SP/STAGING/local_assets/assets/pdfs/Baker_panel_report.pdf).
- CCPS, 2008. "Incidents That Define Process Safety", Center for Chemical Process Safety, New York 2008.
- CCPS, Process Safety Beacon, Facility Siting, March 2010 (http://sache.org/beacon/products.asp)
- CCPS, Process Safety Beacon, Instrumentation – Can You Be Fooled By It?, (http://sache.org/beacon/products.asp)

3.2 Asset Integrity and Reliability: ARCO Channelview, Texas Explosion, 1990

3.2.1 Summary

A wastewater tank containing process wastewater with hydrocarbons and peroxides, exploded during the restart of an off gas compressor. The normal nitrogen purge had been reduced during the maintenance period, and a temporary oxygen analyzer failed to detect the buildup of a flammable atmosphere. When the compressor was restarted, a flammable mixture of hydrocarbons and oxygen were pulled in and ignited. The flashback of the flame into the headspace of the tank ignited the confined vapors and an explosion occurred. The explosion killed 17 people. Damages were estimated to be $100 million.

3.2.2 Detailed Description

The 900,000 gallon wastewater tank contained process wastewater from propylene oxide and styrene processes. Peroxide and caustic by-products from these processes traveled through thousands of feet of piping to the tank where peroxides and caustic mixed. There was normally a layer of hydrocarbons on the surface of the water. A nitrogen purge was used to keep the vapor space inert, and an off-gas compressor drew the hydrocarbon vapors off before the waste layer was disposed of in a deep well. Figure 3.5 shows the process scheme.

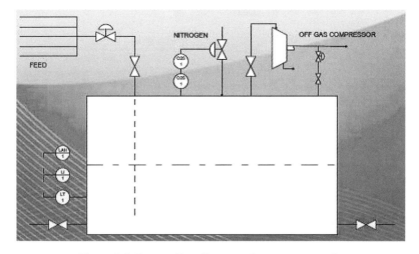

Figure 3.5. Process flow diagram of wastewater tank.

The tank was taken out of service to repair the nitrogen blanket compressor. A temporary oxygen analyzer was installed between two roof beams and provisions made to add a nitrogen purge if a high oxygen level was detected. During this time, the oxygen analyzer failed, giving incorrect low readings. The normal flow of nitrogen purge gas to the tank was reduced. About 34 hours before the explosion, the nitrogen sweep stopped. Therefore, the nitrogen purge was inadequate to prevent a flammable atmosphere from being formed in the headspace and in piping to the compressor. When the compressor was restarted, flammable vapors were drawn in and ignited. Flames flashed back to the tank, causing an explosion in the head space.

When the unit was rebuilt, the new wastewater tank was pressurized and vent gas was sent to a flare. Redundant oxygen analyzers were installed, a backup supply of nitrogen was provided. The preventive maintenance program for the oxygen analyzers and other safety critical equipment was improved. Critical process safety operating parameters were identified for continuous monitoring.

3.2.3 Causes

The oxygen analyzer failed, and the loss of nitrogen sweep was not noticed by the operators.

3.2.4 Key Lessons

Asset Integrity and Reliability (Section 2.11). Safety critical equipment needs to be identified and a preventative maintenance program should be in place to test such equipment.

Conduct of Operations (Section 2.16). Safe operating parameters need to be identified and monitored by operating personnel. It is as important to understand and manage the risks of auxiliary operations, such as this wastewater tank, at the same level as the rest of the process. The use of one oxygen analyzer created a safety critical system that had one point of failure. In designing safety systems, engineers should consider the level of reliability of safety critical systems and provide the necessary redundancy.

3.2.5 References and Links to Investigation Reports

- A Briefing on the ARCO Chemical Channelview Plant July 5, 1990 Incident, ARCO Chemical Company, January 1991.

3.3 Process Safety Culture: NASA Space Shuttle Columbia Disaster, 2003

3.3.1 Summary

The NASA Space Shuttle, Columbia, was destroyed during its re-entry into the Earth's atmosphere at the end of a 16-day voyage, just 16 minutes before scheduled touchdown. During the launch, a large piece of insulation foam became detached from the area where the shuttle had been attached to the external fuel tank and hit the leading edge of the left wing. After the incident, it was discovered that a fragment of the thermal protective panel drifted away from the wing while in space. At the critical part of re-entry when friction with the Earth's atmosphere is at its greatest, superheated air entered the left wing, destroying the structure and causing the spacecraft to lose aerodynamic control, and break up (Figure 3.6). All seven of the crew were killed. Within two hours of loss of signal from Columbia, the independent Columbia Accident Investigation Board (CAIB) was established following procedures that had been put in place after the Challenger disaster 17 years earlier. (CCPS, 2008)

3.3.2 Detailed Description

Columbia was launched on January 16, 2003 for the 28th time. At 81.7 seconds into the flight, a large piece of insulation foam became detached. The detached piece of foam hit the leading edge of the left wing 0.2 seconds later (Figure 3.7).

This event was not detected by the crew or ground support functions until detailed examination of the launch photographs and videos took place the following day. There was sufficient concern that a Debris Assessment Team was created to determine whether the event had caused critical damage to the shuttle. No adverse effects were noticed by the crew or support staff as the mission continued. What they did not know was that on the second day of the flight, an object drifted away from the shuttle. The radar signature of this object, discovered after the incident, was consistent with it being a 140 square inch (900 cm^2) fragment of the protective panel from the left wing of the shuttle. At the critical part of re-entry when friction with the Earth's atmosphere is at its greatest, superheated air entered the left wing, destroying the structure and causing the spacecraft to lose aerodynamic control leading to break up.

NEED FOR PROCESS SAFETY 67

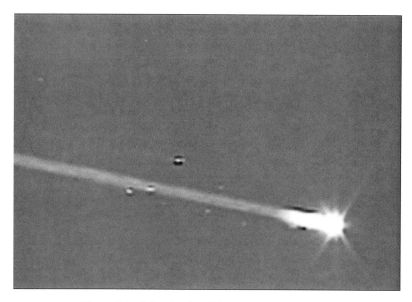

Figure 3.6. Columbia breaking up, courtesy NASA.

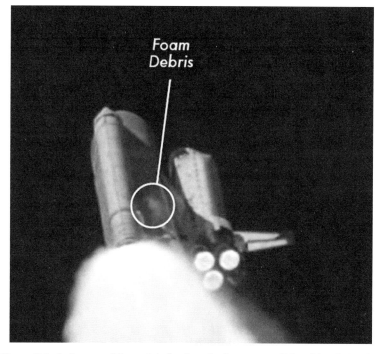

Figure 3.7. A shower of foam debris after the impact on Columbia's left wing. The event was not observed in real time, courtesy NASA.

The obvious question was asked as to how a piece of lightweight foam material could fatally damage something as apparently strong as a spacecraft designed for one of the most aggressive of operating environments. Calculations showed that, at the time of separation, the foam was traveling at the same speed as Columbia - about 1,568 mph (700 m/s) and the rapid deceleration of the foam combined with continued acceleration of the shuttle explained the severity of the impact. It was also found that insulation foam loss had occurred on all previous Space Shuttle flights, from small "popcorn" sized pieces, to briefcase sized chunks, and that, on 10% of flights, foam loss had occurred at the bipod attachment area. In the original design team there had been extreme concern that foam loss would result in fatal damage to the shuttle. Since the specification for the large external fuel tank contained a requirement that "no debris shall emanate from the critical zone of the external tank on the launch pad or during ascent", no protection had been provided to the leading edges of the shuttle's wings. Despite this, there had been a lot of damage to Columbia's protective tiles during its first mission – more than 300 had to be replaced. One engineer stated that if they had known in advance the extent of the debris shower that occurred, they would have had difficulty in getting the Space Shuttle cleared for flight.

Over the previous decade, NASA was placed under severe pressure to reduce costs, losing about 40% of its budget and workforce. Part of the response was for NASA to hand over much of its operational responsibilities to a single contractor, replacing its direct involvement in safety issues with a more indirect performance monitoring role. NASA managers continued to preach the importance of safety, but their actions sent the opposite signal.

Despite the cutbacks, there was pressure — some self-imposed — to keep the Space Shuttle program on schedule, particularly to complete the International Space Station (ISS). The uncertainty over the long-term future of the program resulted in reduced investment, with safety upgrades delayed or deferred. The CAIB found that the infrastructure had been allowed to deteriorate, and the program was operating too close to too many margins.

3.3.3 Causes

Technical. Technically, the cause was the failure of the foam insulation at the bipod attachment. No non-destructive testing (NDT) of hand applied foam was carried out other than visual inspection at the assembly building and at the space center, even though NDT techniques for foam had been successfully used elsewhere. The CAIB concluded that too little effort had gone into the understanding of foam fabrication and failure modes.

Culture. In spite of cutbacks and deadline pressures, the organization continued to pride itself in its "can do" attitude, which had undoubtedly contributed to former successes. This enabled the phenomenon known as "normalization of deviation". The failure of the foam without significant consequences was observed so many times that it became an accepted part of every flight and with each successful landing the original concerns seem to have faded away.

In the words of the CAIB report:

"Cultural traits and organizational practices detrimental to safety were allowed to develop, including: reliance on past success as a substitute for sound engineering practices (such as testing to understand why systems were not performing in accordance with requirements); organizational barriers that prevented effective communication of critical safety information and stifled professional differences of opinion; lack of integrated management across program elements; and the evolution of an informal chain of command and decision-making processes that stifled professional differences of opinion; and decision-making processes that operated outside the organization's rules." (CAIB, 2003).

3.3.4 Key Lessons

Process Safety Culture (Section 2.2). This element was also a key component in the BP Texas City incident (Section 3.1). An important aspect of a good safety culture is maintaining a sense of vulnerability. An example of the poor safety culture at NASA is the denial of requests by the Debris Assessment Team for imaging of the wing while the shuttle was in orbit. The team concluded, based on modeling that "some localized heating damage would most likely occur during re-entry, but they could not definitively state that structural damage would result." The Mission Management Team eventually concluded the debris strike was a "turnaround" issue. As stated in the CAIB report "Organizations that deal with high-risk operations must always have a healthy fear of failure – operations must be proved safe, rather than the other way around. NASA inverted this burden of proof."

The following is a finding from the CAIB report:

- "NASA's safety culture has become reactive, complacent, and dominated by unjustified optimism. Over time, slowly and unintentionally, independent checks and balances intended to increase safety have been eroded in favor of detailed processes that produce massive amounts of data and unwarranted consensus, but little effective communication. Organizations that successfully deal with high-risk technologies create and sustain a disciplined safety system capable of identifying, analyzing, and controlling hazards throughout a technology's life cycle."

3.3.5 References and Links to Investigation Reports

The CAIB report is available online. The online report also contains several movie clips, such as the actual foam strike and impact testing.

- Columbia Accident Investigation Board, (2003) Volume 1
  http://www.nasa.gov/columbia/caib/html/start.html
- Columbia Incident Investigation Board, (2003) Volume 1, movie clips
  http://www.nasa.gov/columbia/caib/html/movies.html
- CCPS, "Incidents That Define Process Safety", Center for Chemical Process Safety, New York, 2008.
- CCPS, Process Safety Beacon, Process Safety Culture, June 2007. (http://sache.org/beacon/files/2007/06/en/read/2007-06-Beacon-s.pdf)

3.4 Process Knowledge Management: Concept Sciences Explosion, Hanover Township PA, 1999

3.4.1 Summary

"At 8:14 pm on February 19, 1999, a process vessel containing several hundred pounds of hydroxylamine (HA) exploded at the Concept Sciences, Inc. (CSI), production facility near Allentown, Pennsylvania. Employees were distilling an aqueous solution of HA and potassium sulfate, the first commercial batch to be processed at CSI's new facility. After the distillation process was shut down, the HA in the process tank and associated piping explosively decomposed, most likely due to high concentration and temperature.

Four CSI employees and a manager of an adjacent business were killed. Two CSI employees survived the blast with moderate-to-serious injuries. Four people in nearby buildings were injured. Six firefighters and two security guards suffered minor injuries during emergency response efforts.

The production facility was extensively damaged (Figure 3.8). The explosion also caused significant damage to other buildings in the Lehigh Valley Industrial Park and shattered windows in several nearby homes." (CSB, 2002).

3.4.2 Detailed Description

Pure HA is a compound with the formula NH_2OH. Solid HA consists of colorless or white crystals that are unstable and susceptible to explosive decomposition and explodes when heated in air above 70°C (158 °F). HA is usually sold as a 50 wt. % or less solution in water. CSI's material safety data sheet (MSDS) stated "Danger of fire and explosion exists as water is removed or evaporated and HA concentration approaches levels in excess of about 70%" (CSI, 1997). HA can be ignited by contact with metals and oxidants.

NEED FOR PROCESS SAFETY

Figure 3.8. Damage to Concept Sciences Hanover Facility, courtesy Tom Volk, The Morning Call.

CSI developed a four step process to make 50% HA:

1. Reaction of HA sulfate and potassium hydroxide to produce a 30 wt.% HA and potassium sulfate aqueous slurry:

   The reaction is:

   $(NH_2OH)_2 * H_2SO_4 \ + \ \ \ 2\ KOH_2 \Rightarrow \ \ NH_2OH \ + \ K_2SO_4 + 2\ H_2O$

   Hydroxylammonium Sulfate   Hydroxylamine

2. Filtration of the slurry to remove precipitated potassium sulfate solids.
3. Vacuum distillation of HA from the 30 wt. % solution to separate it from the dissolved potassium sulfate and produce a 50 wt. % HA distillate.
4. Purification of the distillate through ion exchange cylinders.

The distillation process is shown in Figure 3.9. The charge tank was a 2,500 gallon (9.5 m3) tank. In the first step of the distillation, a pump circulated 30 wt. % HA to the heating column, which is a vertical shell and tube heat exchanger. The HA is heated under vacuum by 49 °C (120 °F) water. Vapor was drawn off to the condenser and collected in the forerun tank and concentrated HA was returned to the charge tank. When the concentration in the forerun tank reached 10 wt. %, it was then collected in the final product tank. At the end of the first step of the

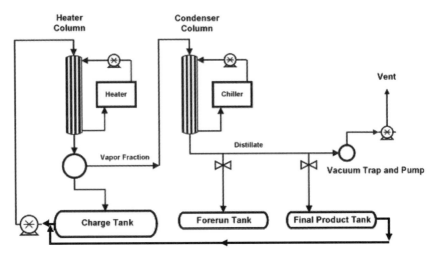

Figure 3.9. Simplified process flow diagram of the CSI HA vacuum distillation process, courtesy CSB.

distillation, the HA in the charge tank was at 80 - 90 wt. % HA. At that point, the charge tank was supposed to be rinsed with 30 wt. % HA.

The first distillation was done by CSI began on Monday afternoon, February 15, 1999. By Tuesday evening, the HA in the charge tank was approximately 48 wt. %. At that time, the process was shut down for maintenance after it was discovered that water had leaked into the charge tank through broken tubes in the heater column. The distillation restarted on Thursday afternoon and shut down at 11:30 PM. The distillation restarted late on Friday morning, after a feed line to the heater column was replaced. By about 7:00 PM, the concentration in the charge tank had reached 86 wt. % HA. The distillation was shut down at 7:15 PM. A manufacturing supervisor was called on Friday evening and arrived at the facility about 15 minutes before the explosion occurred, at 8:14 PM.

It is not known what initiated the explosion. Possibilities include: "addition of excessive heat to the distillation system, physical impacts from partial or total collapse of the glass equipment, or inadvertent introduction of impurities. Friction may have heated the mixture as it passed through the pump that supplied the heating column." (CSB, 2002)

3.4.3 Cause

CSI developed a process for making 50 wt. % HA that would normally cause HA to be concentrated to a level that was inherently unstable and subject to exothermic decomposition.

NEED FOR PROCESS SAFETY

3.4.4 Key Lessons

Process Knowledge Management (Section 2.7). Although management in CSI learned of the hazards of HA during pilot plant operation, the knowledge was not used in the design of the process, or in the hazard reviews that were conducted. CSI also did not review available literature about HA, which also would have shown that HA is subject to exothermic decomposition and had an explosive force equivalent to TNT. Nor did CSI attempt to do any testing to define the magnitude of the hazard of HA. Finally, CSI did not create engineering drawings or detailed operating procedures.

Hazard Identification and Risk Analysis (Section 2.8). Hazard review methodologies need to be appropriate to the hazards being managed. CSI used a "What-If" review which was reported in a one page document. The hazard review did not address the "prevention or consequences of events that could have triggered an explosion of high concentrations of HA". (CSB, 2002). CSI did not implement any of the recommendations from the hazard review they did conduct.

Post-Script. In June 2010 an explosion occurred in a Nissin Chemical Company plant that produced 50 wt. % HA. That explosion destroyed the distillation tower, killed four people and injured 58. The HA concentration at the time was 85 wt. %.

3.4.5 References and links to Investigation Reports

- U.S. Chemical Safety and Hazard Investigation Board, Case Study, Report No. 1999-13-C-PA, The Explosion at Concept Sciences: Hazards of Hydroxylamine, Concept Sciences, Hanover Township, PA. February 19, 1999. (http://www.csb.gov/investigations).

3.5 Hazard Identification and Risk Assessment: Esso Longford Gas Plant Explosion, 1998

3.5.1 Summary

A major explosion and fire occurred at Esso's Longford gas processing site in Victoria, Australia. Two employees were killed and eight others injured. The incident caused the destruction of Plant 1 and shutdown of Plants 2 and 3 at the site.

A process upset in a set of absorbers eventually caused temperature decreases and loss of flow of a "lean oil" stream. This allowed a metal heat exchanger to become extremely cold and brittle. When operators restarted flow of the lean oil to the heat exchanger, it ruptured, releasing a cloud of gas and oil. When the cloud reached an ignition source, the fire flashed back to the release and exploded.

3.5.2 Detailed Description

The plant involved, Plant No. 1, was a lean oil absorption plant, which separated methane from LPG by stripping the incoming gas with a hydrocarbon stream called "lean oil". Methane rises to the top of the towers, with heavier hydrocarbons dissolving in the liquid hydrocarbon condensate, see Figure 3.10.

Plant No. 1 had a pair of absorbers operating in parallel. Each absorber had a gas/liquid disengaging region at the base where a mixture of gas and liquid hydrocarbons entered the absorbers. During the previous night shift, the hydrocarbon condensate level had started to increase in the base of Absorber B. As the normal disposal of condensate to Gas Plant No. 2 was not available, the alternative condensate disposal route was to a Condensate Flash Tank, see Figure 3.11. Under this set of circumstances, it was normal to increase the temperature at the base of the absorber, but this was not done. The inlet to the Condensate Flash Tank was protected against excessively low temperatures by an override on the absorber level controllers. The consequence, therefore, was that the disposal rate of condensate from the absorber became less than the inlet flow, resulting in a buildup of liquid condensate in the absorber base.

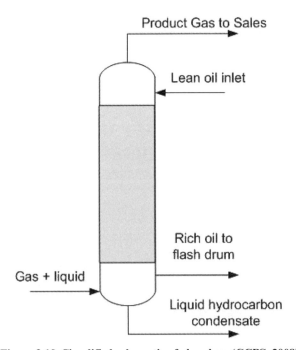

Figure 3.10. Simplified schematic of absorber, (CCPS, 2008).

NEED FOR PROCESS SAFETY

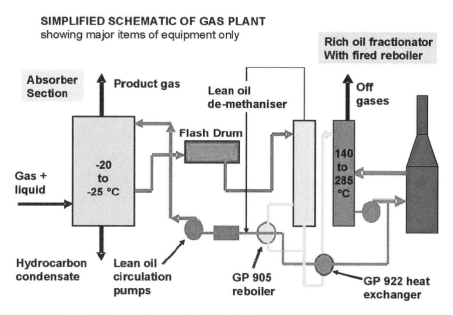

Figure 3.11. Simplified schematic of the gas plant (CCPS, 2008).

The condensate level rose in the absorber to a point where it mixed with the exiting rich stripping oil stream. Condensate mixed with rich oil flashed over the rich oil level control valve resulting in a much reduced temperature in the downstream Rich Oil Flash Tank. This caused temperatures to drop across the plant as rich oil flowed through the recovery process where hydrocarbons were stripped from the rich oil before returning it to the absorbers as lean oil. Eventually, the lean oil pumps tripped out, causing major thermal excursions on a plant with a high degree of process and thermal integration. Loss of lean oil was a critical event, but was not communicated to the supervisor until he returned from the morning production meeting one hour after the pumps had tripped.

Temperatures in parts of the plant fell to -48°C. At 08:30 AM, a condensate leak occurred on heat exchanger GP922. The absence of lean oil flow meant that the condensate flowing through the rich oil system was not warmed as it entered the recovery section. The reason for the leak was probably due an extreme thermal gradient created while attempts were being made to re-establish the process. Other parts of the process showed signs of extreme cold with ice forming on uninsulated parts of heat exchangers and pipework.

At 10:50 AM, the leak from GP922 was getting worse, and the Supervisor decided to shut down Gas Plant No: 1. By 12:15 PM, two maintenance technicians had completed retightening of the bolts on GP922 without making any appreciable difference to the leak. It was decided that the only way to stop the leak was to slowly warm GP922 by starting a flow of warm lean oil through it. However, initial attempts to restart the lean oil pumps were unsuccessful. Ten minutes later, after operating a hand switch to minimize flow through another heat exchanger, GP905, that heat exchanger ruptured, releasing a cloud of gas and oil.

It is estimated that the cloud traveled 170 meters before reaching fired heaters where ignition occurred. After flashing back to the point of release flames impinged on piping, which started to fail within minutes. A large fireball was created when a major pressure vessel failed one hour after the fire had started. It took more than two days to isolate all hydrocarbon streams and finally extinguish the fire (CCPS, 2008).

3.5.3 Cause

The investigation concluded that the immediate cause of the incident was loss of lean oil flow leading to a major reduction in temperature of GP905, resulting in embrittlement of the steel shell. This was followed by introduction of hot lean oil in an attempt to stop the hydrocarbon leak in GP922. Throughout the whole sequence of events, operators and supervisors had not understood the consequences of their actions to re-establish the plant. Esso and the Government were desperate not to shut down the plant, as it supplied all the gas to the State of Victoria. They found their drawings were out of date as they walked the lines to discover what to isolate. In the end they had to shut down the plant and that left the state without power or gas for over ten days, causing major industrial disruption and job losses.

3.5.4 Key Lessons

Hazard Identification and Risk Analysis (Section 2.8). Gas plant #1 had not been subject to a hazard identification study as had been done for the other two gas plants at the site. A Hazard and Operability review, HAZOP, had been planned in 1995, but never carried out. Flow and temperature deviations, like those that occurred at Longford Plant No. 1, are systematically reviewed as part of a HAZOP study. Therefore, the hazardous consequences of these deviations were never identified. This leads to other management safety issues. Procedures and training will be incomplete or inadequate; hence operators will have no knowledge of the seriousness of the deviation. They will not know what to do, and, as in this case, can take the wrong action.

Management of Change (Section 2.14). All of the plant's engineers were relocated to the head office in Melbourne, Australia in 1992. There was no Management of Change review about the effect of removing the process safety tasks that the engineers fulfilled. As a result, their critical roles with respect to process safety were not replaced. The subject of Management of Organizational Change (CCPS, 2013) has been frequently overlooked in the past.

Process Safety Competency (Section 2.4). Supervisors and operators were given greater responsibility for operating the plant, including troubleshooting, for which they were not properly prepared. They were not competent to perform the functions the engineers served.

3.5.5 References and Links to Investigation Reports

- CCPS, "Incidents That Define Process Safety", American Institute of Chemical Engineers, Center for Chemical Process Safety, New York, NY, 2008.
- CCPS 2013, Guidelines for Managing Process Safety Risks During Organizational Change, American Institute of Chemical Engineers, Center for Chemical Process safety, New York, NY, 2013.

3.6 Operating Procedures: Port Neal, IA, Ammonium Nitrate Explosion, 1994

3.6.1 Summary

On December 13, 1994, an explosion occurred in the ammonium nitrate (AN) portion of a fertilizer plant in a process vessel known as a neutralizer. The explosion occurred while the AN process was shut down with AN solution left in several vessels. There were multiple factors contributing to the explosion, including strongly acidic conditions in the neutralizer, application of 200 psig steam to the vessel, and lack of monitoring of the AN plant when the process was shut down with materials left in the process vessels. Four people were killed and 18 injured. There was serious damage to other parts of the plant, resulting in the release of nitric acid to the ground and anhydrous ammonia into the air.

3.6.2 Detailed Description

The Port Neal, IA plant produced nitric acid, ammonia, ammonium nitrate, urea and urea-ammonium nitrate. In the neutralizer, ammonia from the urea plant off gas or from ammonia storage tanks was added through a sparge in the bottom and 55% nitric acid was added through a sparge ring in the middle. The product, 83% AN, was sent to a rundown tank via an overflow line for transfer to storage. See Figure 3.12 for a PFD of the neutralizer and rundown tank. A pH probe was located in the overflow line to the rundown tank and was used to control the nitric

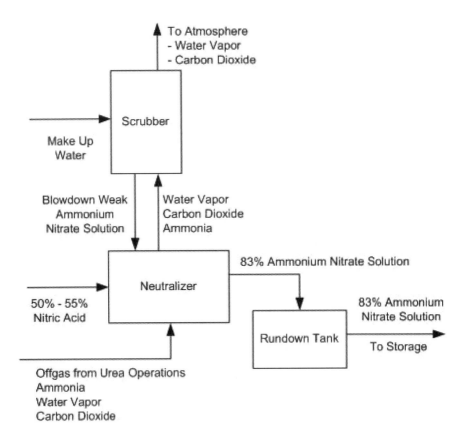

Figure 3.12 Neutralizer and rundown tank, source, (EPA, 1996).

acid flow to the Neutralizer to maintain the pH at 5.5 - 6.5. Temperature in the neutralizer was maintained at about 267 °F (131 °C) by the evaporation of water and ammonia. Both vessels were vented to a scrubber where the vapors were absorbed by 55 - 65% nitric acid and makeup water to make 50% AN. A stream of 50% AN was sent back to the Neutralizer.

About two weeks prior to the event, the pH probe was found to be defective, and the plant was controlled by manually taking samples for pH. Two days prior to the event, the pH was determined to be -1.5 and was not brought into the acceptable range until about 1 AM on December 12. The AN plant was shut down at about 3:00 PM on the afternoon of December 12, because the nitric acid plant was out of service. At about 3:30 PM, operators purged the nitric acid feed line into the neutralizer with air. At about 7:00 PM, operators pumped the scrubber solution to the neutralizer. At about 8:30 PM, 200 psig (13.7 Bar) steam, which is about 387 °F (197 °C), was applied through the nitric acid feed line to the nitric acid sparger to prevent backflow of AN into the nitric acid line. The explosion occurred at about 6 AM on the morning of the 13th. Figure 3.13 shows the aftermath of the explosion.

AN is known to become more sensitive to decomposition, deflagration and detonation by:

- Low pH levels
- High temperatures
- Low density areas (e.g., caused by gas bubbles)
- Physical confinement
- Contaminants such as chlorides and metals.
- Confinement by means of a sufficient mass of AN by itself.

Calculations showed that the nitric acid line clearing would have lowered the pH at the time of the shutdown to about 0.8. The steam sparge was left on for 9 hours; calculations showed that it provided enough heat to raise the solution to its boiling point in about 2 hours. The air and steam sparge created gas bubbles in the solution. Chlorides, carried over from the nitric acid plant, were also found to be present in the AN solution (EPA, 1996).

3.6.3 Causes

The EPA investigation concluded the conditions that led to the explosion occurred due to the lack of operating procedures. There were no procedures on how to put the vessels in a safe state at shutdown, or for monitoring the process vessels during shutdown. There were also no procedures being used to monitor for the presence of chloride salts and/or oil in the reaction mass that could further increase sensitivity of AN to dangerous decomposition conditions.

Figure 3.13. AN plant area after explosion, source, (EPA 1996).

The EPA found that other producers either emptied the process vessels during a shutdown or maintained the pH above 6.0. Also, other producers either did not allow steam sparges or, if steam sparges were done, they conducted them under direct supervision.

The EPA also noted that no hazard analysis had been done on the AN plant, and that personnel interviewed "indicated they were not aware of many of the hazards of ammonium nitrate."

3.6.4 Key Lessons

Operating Procedures (Section 2.9). Operating procedures need to cover all phases of operation. In this event the lack of procedures for shutdown and monitoring the equipment during shutdown led to operators performing actions that sensitized the AN solution, and provided energy to initiate the decomposition reaction.

Hazard Identification and Risk Analysis (Section 2.8). As with the Esso Longford explosion (Section 3.5), a hazard assessment of the AN process had not been done. The lack of a hazard identification study led to personnel not knowing conditions that led to sensitization of AN to decomposition. An effective PHA of the shutdown step would have revealed to the operating staff that the pH of the neutralizer could not be measured with no overflow going into the rundown tank and that the temperature of the neutralizer could not be accurately known without any circulation in the tank. A competently done hazard identification study would have covered backflow of ammonium nitrate into the nitric acid line and better design solutions could have been identified.

3.6.5 References and Links to Investigation Reports

- EPA 1996, Chemical Incident Investigation Report, Terra Industries Inc., Nitrogen Fertilizer Facility, Port Neal, IA, EPA, September 1996.

3.7 Safe Work Practices: Piper Alpha, North Sea, UK, 1988

3.7.1 Summary

An explosion occurred on the Piper Alpha offshore platform, owned by Occidental Petroleum off the Scottish coast of the North Sea. The initial explosion set off a chain of fires and explosions resulting in the loss of 167 lives and near-total destruction of the platform. The explosion began as a release of flammable gas through a poorly installed flange on a line that was improperly put into service. This set off a chain of more explosions and fires that destroyed the platform.

NEED FOR PROCESS SAFETY 81

Figure 3.14. Piper Alpha platform, source (CCPS, 2008).

Sixty-one members of the crew survived the event by jumping into the water and being rescued by boat (Figure 3.14).

The platform layout consisted of a drilling derrick at one end, a processing area in the center, and living accommodation for its crew on the opposite end (Figure 3.15). Piper Alpha acted as a gas gathering facility for two other platforms in the area: Tartan and Claymore "A", receiving high-pressure gas through risers leading from undersea pipelines. Piper Alpha processed the gas from the risers together with its own gas and oil and piped the final products to shore in two separate pipelines (CCPS, 2008).

3.7.2 Detailed Description

The chain of events started when a standby pump was taken off line for maintenance and the valve was nor replaced, and the line was blanked off. Later, a

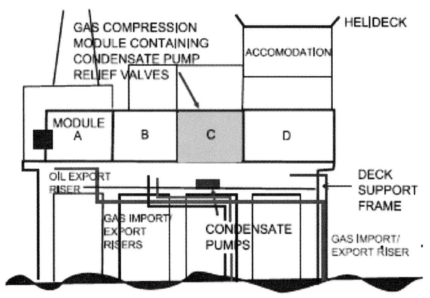

Figure 3.15. Schematic of Piper Alpha platform, source (CCPS, 2008).

condensate pump reinjecting hydrocarbon liquids from the gas/liquid separation process back into the oil export line stopped in the late evening. Attempts to restart it were unsuccessful, and a decision was taken to start up the standby pump as liquid levels were rising rapidly in the process vessels. If not reversed, this would have resulted in total shutdown of the platform. The night shift crew was aware that the standby pump had been taken out of service for maintenance earlier the same day, but believed that the maintenance work had yet to commence. They re-energized the pump motor, which had not been locked out, and started the pump. Within seconds a large quantity of condensate and gas began to escape from the pump discharge pressure relief valve location, in the module above and out of sight of the pump. The relief valve had been removed for maintenance with blind flanges isolating the pipework connections, but with far fewer than the required number of bolts to take full operating pressure.

The condensate pumps were located at the 68 feet Deck Support Frame level, below the modules. The condensate pump relief valves were located inside the Gas Compression Module "C", with the connecting pipework entering and exiting Module "C" through the floor. Module "C" was separated from Module "D" containing the control room and emergency facilities with a non-structural firewall consisting of 3 sheets of a composite plating with mineral wool laid between steel sheets designed to be fire and blast resistant. The fire walls between modules "C" and "B", and "B" and "A" were built from a single plate coated with a fireproofing insulation material. The firewalls installed between the modules were not designed to withstand blast from within any of these modules. An explosion occurred which blew down the firewall containing the processing facility and separating it from the control room. As a result, large quantities of stored oil were almost immediately burning out of control.

The automatic seawater deluge system, which was designed to extinguish such a fire, was unable to be activated as it had been isolated to protect divers carrying out inspection and maintenance on the platform supporting structure near the fire pumps submerged inlets.

About twenty minutes after the initial explosion, the fire had spread to the gas risers generating sufficient heat to cause them to fail catastrophically. The risers were constructed of 24 and 36 inch (610 and 915 mm) diameter steel pipe containing flammable gas at 138 bar-g (2000 psig). When these risers failed, the resulting release of fuel dramatically increased the size of the fire to a towering inferno. At the fire's peak, the flames reached a height of three to four hundred feet. The heat was felt from over a mile away, and reflections in the clouds could be seen from 85 miles.

The crew began to congregate in the platform's living accommodation area, which was the farthest from the blaze and seemed the least dangerous, awaiting helicopters to rescue them. However, the fire prevented helicopters from landing. The accommodation was not smoke-proof and, due to lack of training, people repeatedly opened and shut doors allowing smoke to enter. Some crew members decided that the only way to survive would be to leave the accommodation immediately. However, they found that all routes to lifeboats were blocked by smoke and flames and, lacking any other instructions, they jumped into the sea hoping to be rescued by boat. 61 men survived by jumping. Most of the 167 who died were overcome by carbon monoxide and smoke in the accommodation area. Two men from a rescue vessel died as well.

The gas risers that were fueling the fire were finally shut off about an hour after they had burst, but the fire continued as the oil on the platform and the gas that was already in the pipes burned off. Three hours later the majority of the platform, had burned down to sea level with the derricks and modules, including the accommodation, sliding off and sinking to the sea floor below. Only the drilling part of the platform remained standing above sea level. Oil continued to burn on the sea due to leakage from Piper Alpha's oil production risers (CCPS, 2008).

3.7.3 Causes

The investigation found that the immediate cause of this incident was failure of the Work Permit system to control maintenance and inspection work on the platform. At the beginning of July 6, a Work Permit had been issued for the maintenance of the standby condensate pump. The pump's process connections had been valve isolated and the electrical drive motor isolated and locked off. The first part of the work to be carried out was the removal of the pump's discharge pressure relief valve for inspection. However, only four bolts instead of the full set required for operation were used to fasten the blind flanges fitted over the open ends of the connecting pipework, most likely just to keep the system clean. This pressure relief valve was located in the module above and out of sight of the pump. After removal, the pressure relief valve was taken to the platform workshop for inspection but had not been replaced by the end of the working day.

When the Maintenance Supervisor returned the Work Permit to the Control Room after he and his crew finished their shift, the Process Supervisors and Operators were in deep conversation. Consequently, he left the Work Permit lying on the desk without making any verbal or written hand over. The investigation found that vital communications systems on Piper Alpha had become too relaxed, with the result that the Work Permit was left on the manager's desk instead of it being personally given to him to enable proper communication at the subsequent shift change. If the system had been implemented properly, the initial gas release would not have occurred. However, once this had occurred, many other factors described below conspired together to cause the loss of life and the platform (CCPS, 2008).

3.7.4 Key Lessons

Safe Work Practices (Section 2.10). Good safe work practices are needed to control hazards due to maintenance work. These work practices need to include communication between the people doing the work and production personnel. In Piper Alpha, the night shift crew was not informed that the relief valve had been removed and the pump was not ready to be returned to operation. Additionally, the blind flange put in the line was not properly installed, so it could not hold the pressure in the line.

Emergency Management (Section 2.17). The Offshore Installation Manager (OIM) did not order an evacuation immediately and was killed shortly after. Fire boats responding to the event waited for orders from the OIM, which delayed response. Many of the evacuation routes were blocked. Other wells in the area were feeding material to Piper Alpha and did not turn off their feeds, providing a continuing source of fuel to the fire. The workers on the platform were not adequately trained in emergency procedures, and management was not trained to provide good leadership during a crisis situation. Evacuation drills were performed, but not every week as required by regulations. A full drill had not taken place in over three years. The place where the crew gathered was not safe. Smoke could enter and this caused the fatalities. After Piper Alpha, the UK Government required that there be a Temporary Safe Refuge (TSR) protecting staff sheltering there from explosions, fire and toxic smoke until safe evacuation can be organized.

NEED FOR PROCESS SAFETY

3.7.5 References and Links to Investigation Reports

- CCPS, "Incidents That Define Process Safety", American Institute of Chemical Engineers, Center for Chemical Process Safety, New York, NY, 2008.
- M. Elisabeth Pate-Cornell, Learning from the Piper Alpha Incident: A Postmortem Analysis of Technical and Organizational Factors, Risk Analysis, Vol. 13, No. 2, 1993, p. 215-232.
- CCPS, Process Safety Beacon, Remembering Piper Alpha, July 2013 (http://sache.org/beacon/products.asp)
- CCPS, Process Safety Beacon, Piper Alpha Oil Platform, July 2005 (http://sache.org/beacon/products.asp)

3.8 Contractor Management: Partridge Raleigh Oilfield Explosion, Raleigh, MS, 2006

3.8.1 Summary

"The incident occurred at about 8:30 AM on June 5, 2006, when Stringer's Oilfield Services contract workers were installing pipe from two production tanks to a third, (Figure 3.16). Welding sparks ignited flammable vapor escaping from an open-ended pipe about four feet from the contractors' welding activity on tank 4. The explosion killed three workers who were standing on top of tanks 3 and 4. A fourth worker was seriously injured." (CSB, 2007).

3.8.2 Detailed Description

Contract workers were connecting piping between two recently moved tanks (numbers 3 and 4 in Figure 3.15). Several days before, crude oil residue was removed from tank 4 and it was flushed with water. Crude oil residue was not removed from tanks 2 and 3.

Before starting to weld, the welder checked for flammable vapors in tank 4 by inserting a lit welding torch into it, an act known as "flashing". (Testing for a flammable atmosphere in this manner is an extremely unsafe act, and should never be done.) Then, as the CSB report states "the foreman climbed to the top of tank 4. Two other maintenance workers climbed on top of tank 3; they then laid a ladder on the tank roof, extending it across the 4 foot space between tank 3 and 4 and held the ladder steady for the welder. The welder attached his safety harness to the top of tank 4 and positioned himself on the ladder.

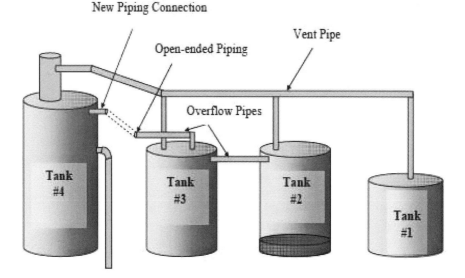

Figure 3.16. Tanks involved in the Partridge Raleigh oilfield explosion, source (CSB, 2006).

Almost immediately after the welder started welding, flammable hydrocarbon vapor venting from the open-ended pipe that was attached to tank 3 ignited. The fire immediately flashed back into tank 3 and spread through the overflow connecting pipe from tank 3 to tank 2, causing tank 2 to explode. The lids of both tanks were blown off." (CSB, 2006).

The tank lid of tank 3 landed 50 feet away, Figure 3.17, and the lid of tank 2 landed about 250 yards away. The foreman and maintenance workers were killed and the welder injured.

3.8.3 Cause

The cause of this incident is conducting hot work in the presence of a flammable atmosphere without following any safe work permitting procedure.

3.8.4 Key Lessons

Safe Work Practices (Section 2.10) Use of safe work practices, such as hot work permits, is necessary to ensure a safe work environment when hazardous chemicals, in this case flammable vapors, are present. The contractor, Stringer's Oilfield Services, did not require the use of safe work procedures, specifically, hot work permits in this case.

Figure 3.17. Tank 3 lid, source (CSB, 2007).

Contractor Management (Section 2.12) Contractors need to be managed in such a way as to ensure they know about and use safe work practices. The owner of the wells and tanks, Partridge-Raleigh, relied on contractors to do most of their well commissioning work, such as installing tanks, pumps, and piping. This is a common practice. Partridge-Raleigh did not, however, manage the contractors to make sure they used safe work practices.

Compliance with Standards (Section 2.3) Companies need to be aware of and follow best industry practices. There are several applicable industry guidelines that covered this situation. If any of these third party standards had been adopted by Raleigh-Partridge or Stringer's Oilfield Services, this incident could have been prevented:

- NFPA 326, "Standard for the Safeguarding of Tanks and Containers for Entry, Cleaning, or Repair" (NFPA, 2005).
- NFPA 51B, "Standard for Fire Prevention During Welding, Cutting, and Other Hot Work" (NFPA, 2003).
- API RP 2009, "Safe Welding, Cutting and Hot Work Practices in the Petroleum and Petrochemical Industries", (API 2002)
- API RP 74, "Occupational Safety for Onshore Oil and Gas Production Operations" (API 2001).

3.8.5 References and Links to Investigation Reports

- U.S. Chemical Safety and Hazard Investigation Board, Case Study, Report No. 2006-07-I-MS, Hot Work Control and Safe Work Practices at Oil and Gas Production Wells. Raleigh, MS, June 5, 2006 (http://www.csb.gov/investigations).
- U.S. Chemical Safety and Hazard Investigation Board, Video – Dangers of Hotwork, June 7, 2010 (http://www.csb.gov/videos).

3.9 Asset Integrity and Reliability: Explosion at Texaco Oil Refinery, Milford Haven, UK, 1994

3.9.1 Summary

A semi-confined vapor cloud explosion occurred on July 24, 1994 on a refinery jointly owned by Texaco and Gulf Oil. Approximately 20 tons of flammable hydrocarbons escaped to atmosphere from the outlet pipe of the flare knock-out drum on the fluidized catalytic cracking unit (FCCU). The drifting cloud of vapor and droplets ignited about 110 meters from the flare drum outlet and the force of the explosion was estimated to be equivalent to 4 tons of high explosive. This was followed by a major fire. The site suffered severe damage with broken glass occurring in a town 2 miles (3 km) away. 26 people suffered injuries on-site, none serious. The costs for rebuilding the damaged refinery were estimated at $76 million and the company was fined $320,000 with $230,000 legal costs. See Figure 3.18 (CCPS, 2008).

3.9.2 Detailed Description

A series of heavy thunderstorms had passed over the area between 7:30 and 9:30 AM that morning. Direct lightning strikes caused a fire in the crude oil distillation unit (CDU), which resulted in a shutdown of the FCCU and some other process units. While attempting to restart the FCCU later that morning, the debutanizer column became starved of feed with the result that the bottoms liquid level control valve automatically closed. The debutanizer later became flooded with hydrocarbon liquid as its feedstock was re-established, most probably because the bottoms liquid level control valve became stuck in the closed position. As the column filled, its internal pressure increased and the debutanizer pressure relief valves (PRV's) opened allowing a mixture of light hydrocarbon liquids and vapors to enter the flare system.

NEED FOR PROCESS SAFETY

Figure 3.18. Ref. (CCPS, 2008) Picture courtesy of Western Mail and Echo Ltd.

Despite the major upset that had occurred on the FCCU gas recovery system, it was decided to restart a wet gas compressor to maintain pressure differentials and thus, hopefully, to re-establish stable operating conditions. In order to do this, it was necessary to drain large quantities of hydrocarbon liquids from the compressor knock out drums. This was done by temporarily connecting steam hoses between the process vessels and the flare header. The wet gas compressor was restarted successfully, but caused the pressure within the debutanizer to rise again resulting in the column PRV's reopening to flare. Liquid levels also continued to rise in the wet gas compressor knock out drums until eventually it tripped due to a high level in its inter-stage knock out drum. By this time, the combined liquid hydrocarbon flows to the flare line knock out pot had filled it well above its design capacity. At 12:32 PM, the 30 inch (760 mm) diameter flare knock out pot outlet pipe ruptured at its weakest point. It was concluded that the sudden hydraulic forces caused by major liquid hydrocarbon flows entering the flare system had caused it to break at an elbow (CCPS, 2008). See Figure 3.19.

3.9.3 Causes

The debutanizer bottom product level control valve had never reopened after it had closed when the column became starved of feed. Operators had received incorrect signals indicating that it had reopened, but the debutanizer continued to fill with liquid while the downstream naphtha splitter remained empty.

Figure 3.19. The 30 inch flare line elbow that failed and released 20 tons of vapor, source (HSE, 1994).

3.9.4 Key Lessons

Asset Integrity and Reliability (Section 2.11). Instrumentation needs to be assessed and maintained in the same manner as process vessels and rotating equipment items. In addition to the level control valve, 39 instruments in the gas recovery system were found to have had pre-existing faults of which 24 required attention and 6 were seriously deficient.

Also, the knock out pot outlet line that failed was severely corroded. According to the HSE, which investigates incidents in the UK, "The presence of corrosion was known, but the full extent was not recognized because it had not been inspected at the point of failure, where there were inspection access difficulties."

3.9.5 References and Links to Investigation Reports

- CCPS (2008), "Incidents That Define Process Safety", American Institute of Chemical Engineers, Center for Chemical Process Safety, New York, NY, 2008.
- HSE (1994) "The Explosion and Fires at Texaco Refinery, Milford-Haven, 24 July 1994.
(https://www.icheme.org/~/media/Documents/Subject%20Groups/Safety_Loss_Prevention/HSE%20Accident%20Reports/The%20Explosion%20and%20Fires%20at%20the%20Texaco%20Refinery%20Milford%20Haven.pdf)
- U.S. Chemical Safety and Hazard Investigation Board, Video – Chevron Richmond Refinery Fire (http://www.csb.gov/chevron-refinery-fire/)
- CCPS Process Safety Beacon, Mechanical Integrity, April 2006 (http://sache.org/beacon/files/2006/04/en/read/2006-04-Beacon-s.pdf)

3.10 Conduct of Operations: Formosa Plastics VCM Explosion, Illiopolis, IL, 2004

3.10.1 Summary

"On April 23, 2004, an explosion and fire killed five and seriously injured three workers at the Formosa Plastics Corporation, IL (Formosa-IL) PVC manufacturing facility in Illiopolis, Illinois. The explosion occurred after a large quantity of highly flammable vinyl chloride monomer (VCM) was inadvertently released from a reactor and ignited. The explosion and fire that followed destroyed much of the facility and burned for two days (Figure 3.20). Local authorities ordered residents within one mile of the facility to evacuate." (CSB, 2007). The damage led to the facility being permanently closed.

3.10.2 Detailed Description

Polyvinyl chloride (PVC) was made by heating vinyl chloride monomer (VCM), water, suspending agents, and polymerization reaction initiators under pressure in a batch reactor. VCM is a highly flammable material. There were 24 reactors in a building, and the reactors were put in groups of 4 with a control station for every two reactors (Figure 3.21). When a reaction was complete, the PVC solution was transferred through the bottom valve to a vessel for the next step in the process.

After the transfer, the reactor was purged of hazardous gases and cleaned by power washing it through an open manway. The wash water was emptied to a drain through the reactor bottom valve and a drain valve. All of these steps were done manually.

Figure 3.20. Smoke plumes from Formosa plant, source (CSB 2007).

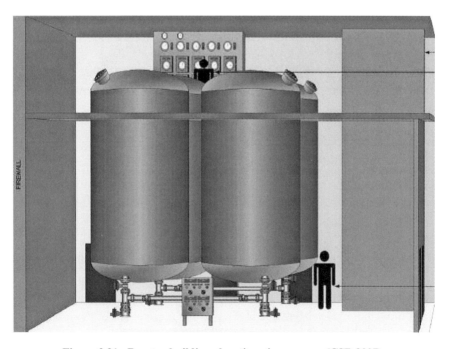

Figure 3.21. Reactor building elevation view, source (CSB 2007).

NEED FOR PROCESS SAFETY

On the day of the incident, the transfer of the reaction mass and the power washing had been completed in one reactor, 306. The blaster operator went downstairs to drain the reactor. At the bottom of the stairway, he turned in the wrong direction towards an identical set of four reactors that were in the reaction phase of the process. See Figure 3.22. The operator mistakenly attempted to empty reactor 310 by opening the bottom and drain valves. The bottom valve, however, was interlocked to remain closed when the reactor pressure was above 10 psi, that is, when it was processing a batch of PVC. Consequently, it did not open. The air supply to the bottom valve had been equipped with quick disconnect fittings and a separate supply of emergency air was provided to allow operators to transfer a batch from one reactor to another in an emergency.

When the bottom valve did not open, the blaster operator switched to the backup air supply thereby overriding the interlock. This was done without contacting the upstairs reactor operator or shift foreman to check on the status of the reactor.

As the bottom valve was opened, VCM poured out of the reactor and the building rapidly filled with flammable liquid and vapor. A deluge system in the building alarmed, but failed to activate (the deluge system might not have prevented the explosion even if it did activate). A shift supervisor came to the area to investigate. There were VCM detectors in the building, and they were reading above their maximum measurable levels. The shift foreman and reactor operators took measures to slow the release rather than evacuate. The VCM vapors found an ignition source and several explosions occurred.

Figure 3.22. Cutaway of the reactor building, source (CSB 2007).

3.10.3 Causes

The operator overrode an interlock without consulting with shift supervision, leading to a release of hot, pressurized VCM. There were several factors that made this error more likely to occur:

- The similar layout of the reactor groupings (see **Figure** 3.21) making it error prone.
- The operators on the lower levels were not given radios to make communication with the reactor control operators on the upper level easier. (Similar Formosa plants had radios or an intercom system.)
- Formosa eliminated an operator group leader position and shifted their responsibility to the shift supervisors, who were not always as available as the group leaders used to be. This reduced the amount of support available to operators.

3.10.4 Key Lessons

Conduct of Operations (Section 2.16). Conduct of Operations means doing tasks in a disciplined manner. That means following operating procedures and protocols, and, when the process moves outside the operating envelope, to stop work, think about the response, get experienced advice as needed, and shut down as appropriate. In the Formosa event, the blaster operator was supposed to get supervisory approval to override the interlock.

Conduct of Operations does not apply to just operators on the process floor; it also applies to the design of controls and systems needed to support operators. Conduct of Operations includes engineering discipline. In the case of the Formosa VCM explosion, the blaster operators had to cope with an error prone design. The reactor layout made it easier for a mix-up to occur, a common practice in engineering design. An emergency transfer procedure required bypassing the bottom valve interlock, so an easy means was provided to do this. Engineers who design and run plants should try to provide engineering controls and monitor shift notes and logs for instances of bypassing interlocks. In this case, a reactor status indication on the operating floor could have been provided and a more rigorous enforcement of operating procedures and interlock management. Operators were not provided tools (radios for communication for operators between floors) to make it easier for them to follow their procedures. It is the responsibility of management to provide the tools and controls necessary for operators to do their jobs safely.

Management of Change (Section 2.14). When Formosa Plastics took over the plant, a reduction in staff and changes in responsibilities were made. There was no formal MOC review of the staffing changes to analyze the impact of these changes.

NEED FOR PROCESS SAFETY

Emergency Response (Section 2.17). The Formosa Plastics VCM explosion also illustrates the importance of emergency response planning. When the VCM release occurred, gas detectors in the building activated and operators responded by trying to mitigate the release. The proper response to this activation should have been to evacuate.

Measurement and Metrics (Section 2.19). There had been two previous incidents at Formosa Plastics plants that involve operators opening the wrong valves. These incidents and their lessons were not shared with the other plants. A metrics system that tracked leading and lagging indicators, as described in Section 2.19, could have alerted Formosa Plastics to a systemic problem and enabled the company to take steps to correct it. These actions could have been to train operators at all plants about why bypassing the interlocks was dangerous and/or placing a guard over the hose connection to prevent overriding the interlock without proper review and authorization.

3.10.5 References and Links to Investigation Reports

- CSB 2007, U.S. Chemical Safety and Hazard Investigation Board, Investigation Report, Vinyl Chloride Monomer Explosion, Report No. 2004-10-I-IL, March 2007.
- U.S. Chemical Safety and Hazard Investigation Board, Video – Explosion at Formosa Plastics, 2007, (http://www.csb.gov/formosa-plastics-vinyl-chloride-explosion/)

3.11 Management of Change: Flixborough Explosion, UK, 1974

3.11.1 Summary

On the evening of June 1 1974, the Nypro (UK) site at Flixborough was severely damaged by a large explosion. 28 employees were killed and 36 injured. The site office building was demolished, but fortunately it was empty because the incident occurred on the weekend. Outside of the plant, injuries and damage were widespread, with 53 people being reported injured. There were no offsite fatalities. Varying degrees of damage was caused to 1,821 houses and 167 business premises (CCPS, 2008).

3.11.2 Detailed Description

The part of the plant on which the explosion occurred involved the oxidation of cyclohexane to cyclohexanone using air injection at 8.8 bar g (130 psig) and 155°C (311°F). The oxidation reaction was carried out in the liquid phase in six reactor vessels arranged in series, each set 360 mm (14 inches) below its predecessor to allow the flow to progress through the reaction train by gravity.

On March 27, 1974, a crack was detected on Reactor No.5. A Maintenance Engineer recommended complete closure for 3 weeks. The Maintenance Manager, whose job had been filled for several months by the head of the laboratory while awaiting a reorganization of the company, proposed dismantling Reactor No. 5 and connecting numbers 4 and 6 together by a 500 mm (20 inch) diameter temporary connection. To support the piping, the proposal was to use a structure made from conventional construction industry scaffolding, Figure 3.23.

The temporary connection was not adequate for the forces and temperatures involved, and failed, releasing 30 metric tons of cyclohexane in 30 seconds. Of the 28 employees killed, 18 were in the control room. The loss of life could have been far higher if the incident had occurred on a weekday, and not on a Saturday, when the number of day employees on site was low. The whole plant was destroyed. The neighboring housing was devastated. The fire lasted over three days with 40,000 m^2 (10 acres) affected. See Figures 3.24 through 3.26 (CCPS, 2008).

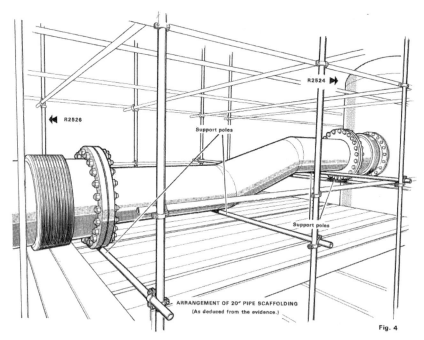

Figure 3.23. Schematic of Flixborough piping replacement, source Report of the Court of Inquiry.

NEED FOR PROCESS SAFETY

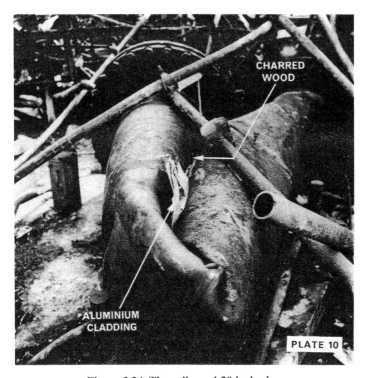

Figure 3.24. The collapsed 20 inch pipe.

Figure 3.25. Damage to Flixborough plant.

Figure 3.26. Damage to Flixborough control room.

3.11.3 Cause

The design of the replacement pipeline was inadequate, leading to its failure. The investigation found that "No consideration was given to the bending moments or hydraulic thrusts that would be imposed on the assembly due to its dogleg design. There was no reference made to vendor manuals for the expansion bellows, nor to relevant British Standards. No drawing was made for the design." (Source, Court of Inquiry, 1975).

3.11.4 Key Lessons

Management of Change (Section 2.14). Changes to a process or equipment must be reviewed and implemented by people with knowledge appropriate to the situation. This incident is important in the history of process safety as the prime example of the importance of an MOC program. There was no engineering review of this change at all. As seen in the cause section, important mechanical design features were not considered during the change.

Flixborough highlights the importance of Management of Organizational Change (MOOC) as well as physical change. At Flixborough, "the works engineer had left early in the year and had not yet been replaced. At the time the bypass line was being planned and installed, there was no engineer on site with the qualifications to perform a proper mechanical design, or to provide critical technical review on related issues. There were chemical and electrical engineers

NEED FOR PROCESS SAFETY 99

on staff, but no other mechanical engineers." A statement often used in relation to the modifications at Flixborough is that "they didn't know what they didn't know". Although the presence of a mechanical engineer may not have changed the outcome if no MOC review was held at all, it is more likely that the significance of the change could have been recognized by someone at the plant. MOOC covers modification of working conditions, personnel turnover, task allocation changes, organizational hierarchy changes, and organizational policy changes. *Guidelines for Managing Process Safety Risks During Organizational Change* (CCPS 2013) covers this topic in more detail.

Codes and Standards (Section 2.3). As stated in the summary, the site office building was destroyed. At the time, 1974, there were no standards in place with respect to facility siting and layout. This event is an example of why there is now such a standard, API RP 752, Management of Hazards Associated with Location of Process Plant Buildings.

3.11.5 References and Links to Investigation Reports

- CCPS 2008, "Incidents That Define Process Safety", American Institute of Chemical Engineers, New York, NY, 2008.
- CCPS 2013, "Guidelines for Managing Process Safety Risks During Organizational Change, American Institute of Chemical Engineers, New York, NY, 2013.
- CCPS Process Safety Beacon, Flixborough – 30 Years Ago, June 2004 (http://sache.org/beacon/files/2004/06/en/read/2004-06%20Beacon-s.pdf)
- CCPS Topics, Incident Summary: Flixborough Case History (http://www.aiche.org/ccps/topics/elements-process-safety/commitment-process-safety/process-safety-culture/flixborough-case-history)
- Her Majesties Stationary Office, The Flixborough Disaster – Report of the Court of Inquiry, 1975.
- API RP 752, Management of Hazards Associated With Location of Process Plant Buildings, 3rd Edition, American Petroleum Institute, December 2009.

3.12 Emergency Management: Sandoz Warehouse Fire, Switzerland, 1986

3.12.1 Summary

On November 1, 1986, a fire broke out at a Sandoz storehouse near Basel, Switzerland. The fire occurred in an unsprinklered warehouse 90 m long by 50 m wide and 8 m high (300 by 165 and 26 ft.), storing chemicals in a high-piled configuration. There were about 1,350 metric tons of at least 90 different chemicals, including insecticides, herbicides, mercury-containing pesticides and phosphoric acid esters. Flames were shooting from the roof when the fire was first noticed. Steel drums of chemicals exploded like bombs in the intense heat (Figure 3.27).

The majority of the stored chemicals were destroyed in the fire, but large quantities were introduced into the atmosphere, into the Rhine River through runoff of firefighting water, and into the soil and groundwater at the site. Nearly 400 firefighters used massive amounts of water to completely extinguish the fire as quickly as possible. About 10,000 to 15,000 m^3 of water was used when the water treatment system could only contain 50 m^3.

The exact mass of chemicals entering the Rhine has been estimated at somewhere between 13 and 30 tons. A toxic red chemical slick 40 km (25 miles) long was created, descending the river at 3 km per hour, resulting in widespread destruction of aquatic life, which only began recovering more than a year following the incident, Figure 3.28.

It was later discovered that nearly all of the water used in fire suppression flowed through storm drained directly into the Rhine. Sandoz had to reimburse the countries affected. The cause of the fire has not been positively determined.

Figure 3.27. Sandoz Warehouse firefighting efforts, source (CCPS, 2008)

NEED FOR PROCESS SAFETY

Figure 3.28. Impact of Sandoz Warehouse firewater runoff, (CCPS, 2008).

3.12.2 Key Lessons

Emergency Management (Section 2.17). It is important to plan for major incidents with all parties that will be involved. Once an event starts, the responses to it will be reactive, and based on perceptions that are potentially wrong, unless it is planned for ahead of time. In the case of a warehouse fire, it is not surprising that the local fire department will respond. In this case, failure to plan for the firefighting efforts led to a major environmental catastrophe.

Similar incident: At a warehouse fire in West Helena, Arkansas in 1997, three firefighters were killed and 16 people injured when an explosion occurred. A plan for emergency response could have advised firefighters to stay away from buildings containing materials that could explosively decompose.

3.12.3 References and links to investigation reports

- CCPS 2008, "Incidents That Define Process Safety", American Institute of Chemical Engineers, Center for Chemical Process Safety, New York, NY, 2008.

3.13 Conduct of Operations: Exxon Valdez, Alaska, 1989

3.13.1 Summary

On March 24, 1989, the Exxon Valdez ran aground on a reef in the Prince William Sound, off the coast of Alaska, at 12:04 AM. See Figure 3.29.

Approximately 11 million gallons or 257,000 barrels of oil was spilled. Approximately 1,300 miles of shoreline were impacted, 200 miles of it heavily or moderately. Exxon spent $2.1 billion on cleanup costs (Exxon Valdez Oil Spill Trustee Council).

3.13.2 Detailed Description

At 11:25 PM, the state pilot (who guides the ship out of the harbor) left the ship and the captain informed the Vessel Traffic Center that he was increasing to sea speed. He also reported that the Exxon Valdez would divert from the outbound lane and end up in the inbound lane if there was no conflicting traffic due to icebergs. The traffic center indicated concurrence, stating there was no reported traffic in the inbound lane. The ship actually went beyond the inbound lane.

At 11:52 PM, the command was given to place the ship's engine on "load program up" a computer program that, over a span of 43 minutes, would increase engine speed from 55 RPM to sea speed full ahead at 78.7 RPM. After conferring with the helmsman about where and how to return the ship to its designated traffic lane, the captain left the bridge. At about midnight, the ship struck the reef. The grounding was described by the helmsman as "a bumpy ride" and by the third mate as six "very sharp jolts". Eight of 11 cargo tanks were punctured. Computations aboard the *Exxon Valdez* showed that 5.8 million gallons had gushed out of the tanker in the first three and quarter hours." (Exxon Valdez Oil Spill Trustee Council). Figures 3.30, 31 and 32 are pictures of the cleanup.

NEED FOR PROCESS SAFETY 103

Figure 3.29. Exxon Valdez tanker leaking oil, courtesy of Exxon Valdez Oil Spill Trustee Council.

Figure 3.30. Oiled loon onshore, courtesy of Exxon Valdez Oil Spill Trustee Council.

Figure 3.31. Aerial of a maxi-barge with water tanks and spill works hosing a beach, Prince William Sound, courtesy of Exxon Valdez Oil Spill Trustee Council.

Figure 3.32. Cleanup workers spray oiled rocks with high pressure hoses, courtesy of Exxon Valdez Oil Spill Trustee Council.

NEED FOR PROCESS SAFETY

3.13.3 Causes

The common view is that human error was the main cause for this event. In fact, there are several causes beyond simply "human error". The National Transportation Safety Board investigated the incident and determined that the probable causes of the grounding were:

1. The failure of Exxon Shipping Company to supervise the master and provide a rested and sufficient crew for the *Exxon Valdez*;
   - Note: The average size of an oil tanker crew was 40 in 1977, the Exxon Valdez had a crew of 19
   - The crews routinely worked 12-14 hour shifts
   - The crew rushed to get the tanker loaded and out of port
2. The failure of the U.S. Coast Guard to provide an effective vessel traffic system
   - Note: The radar station in Valdez had replaced its radar with a less powerful one, the location of tankers near Bligh reef could not be monitored with this equipment.
3. The lack of effective pilot and escort services.
   - Note: The practice of tracking ships out to the Bligh reef had been discontinued; tanker crews were never informed of his.

1. All notes come from Leveson, 2005.

3.13.4 Key Lessons

Conduct of Operations (Section 2.16). Conduct of operations was also mentioned in the Formosa Plastics explosion in Section 3.10. In the Exxon Valdez grounding, several operational requirements for operating the vessel, noted above, were not followed.

Another conduct of operation issue was the failure of the Exxon Shipping Company to supervise the master. When you start working in a petrochemical or other processing plant you may hear about "management walk-arounds". Plant management needs to spend some time in the field observing the conditions of the plant and behavior of personnel and communicating with them. It is through such activities that a manager can observe whether or not the most up to date set of instructions being followed, as in the BP Texas City explosion (Section 3.1). In this case, whether or not the crew was sufficiently rested, and the officers following their procedures were questions of Conduct of Operations.

Management of change (Section 2.14). Causes 4 and 5 show that two major changes were made in the way ships were guided out of Prince William Sound; the on shore radar was downgraded and the practice of tracking ships out of the sound was discontinued. It was noted that ship crews were not informed of the end of that practice. A management of change review, which included a representative of the shipping company, could have either prevented the change or allowed ships to adjust their operations accordingly.

3.13.5 References and Links to Investigation Reports

- Exxon Valdez Oil Spill Trustee Council website, http://www.evostc.state.ak.us/
- Leveson 2005, Leveson, Nancy G, "Software System Safety", July 2005 (http://web.archive.org/web/20101108055426/http://ocw.mit.edu/courses/aeronautics-and-astronautics/16-358j-system-safety-spring-2005/lecture-notes/class_notes.pdf)

3.14 Compliance with Standards: Mexico City, PEMEX LPG Terminal, 1984

3.14.1 Summary

On November 19, 1984, a Liquefied Petroleum Gas (LPG) release in a distribution terminal ignited, and the resulting fires led to the explosion of a series of the LPG storage tanks. About 600 people were killed, around 7,000 injured, 200,000 people were evacuated and the terminal destroyed. Many of those killed lived in simple brick or wooden houses constructed after the terminal had been built, the nearest of these being only 130 meters (426 feet) from the LPG tanks.

3.14.2 Detailed Description

The LPG pressure storage consisted of four spheres of 1600 m^3 and two spheres of 2,400 m^3 capacity, plus 48 horizontal cylindrical storage tanks of various capacities, as shown in Figure 3.33. All storage was contained within 1 meter high concrete walled enclosures. The storage terminal received LPG through three underground pipelines from remote refineries several hundreds of kilometers away. The terminal distributed LPG to local gas companies through underground pipelines, by loading road and rail cars, and as bottled gas from an onsite bottling plant. The site also contained two ground level flare pits, and a fire protection system complete with pond, firewater pumps and water spray systems.

The cause of the release has never been definitely established. One source (Lees, 1996) states that a drop in pressure was noticed in the control room and also at a pipeline pumping station. It may have been caused by the rupture of an 8-inch pipe between a sphere and a series of tanks. Unfortunately, the operators could not identify the cause of the pressure drop. Another source (CCPS, 2008) states that the loss of containment could have come from came from overpressure of a pipeline (in line with Lees). It also states the loss of containment could have come

NEED FOR PROCESS SAFETY

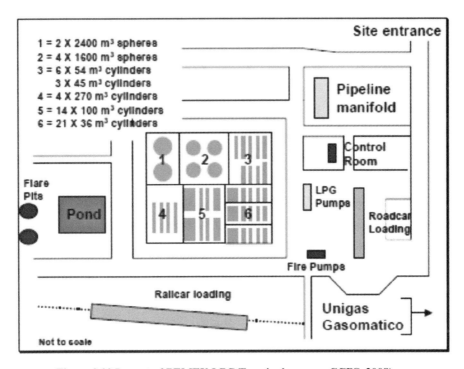

Figure 3.33 Layout of PEMEX LPG Terminal, source, CCPS, 2008).

from overfilling a storage tank. The CCPS source states that the horizontal tanks were undersized so that the flow to them had to be diverted to another tank every 30 - 45 minutes to prevent overflow.

The release of LPG continued for about 5-10 minutes when the gas cloud, estimated at 200 m x 150 m x 2 m high (660 ft. x 490 ft. x 6 ft.), drifted to a flare stack. It ignited, causing violent ground shock. A number of ground fires occurred. Workers in the plant now tried to deal with the release by taking various actions. At a late stage, somebody pressed the emergency shutdown button, with flow to the terminal continuing for one hour after the explosion.

The degree of confinement in the horizontal storage tank enclosure was such that tanks were thrown off their supports and piping ruptured. Nine explosions and BLEVE's followed. The four small spheres were completely destroyed with fragments scattered around the area, some as far as 350 meters (1150 ft.) away in public areas. The spheres collapsed onto the ground as the legs buckled due to the heat of the fire. Only four of the horizontal cylindrical tanks survived, with twelve of them being ejected over 100 meters (330 ft.) from their supports, with the furthest "rocketing" 1200 meters (3940 ft.). Gas entering buildings inside the terminal and the public housing ignited resulting in overpressure explosions. Before and after pictures of the terminal are shown in Figures 3.34 and 3.35.

Figure 3.34. PEMEX LPG Terminal prior to explosion source, CCPS, 2008.

Figure 3.35. PEMEX LPG Terminal after the explosion source, CCPS, 2008.

3.14.3 Causes

Due to the level of destruction and the fact that most of the PEMEX personnel were killed as a result of the fires and explosions, a cause for the initial release cannot be definitely established. Contributing factors were the lack of passive fire protection systems capable of increasing the survivability of critical systems in a major fire situation, and the destruction of the fixed fire protection systems. There are some suggestions, however, that the installed spray water systems were inadequate to protect the storage vessels in any event.

3.14.4 Key Lessons

Compliance with standards (Section 2.3). Facility siting and layout standards, in particular, were not used in the design of this terminal. For example, the LPG vessels were closely spaced. Providing more land area to better space vessels and permit good drainage and LPG spill containment could have reduced the consequences of the failure (e.g., less chance of BLEVE and reduced amount of LPG released). Better access could have permitted a better chance of controlling the fire and containing the release (Lees, 1996).

Also, note that in Figure 3.34, some of the horizontal storage tank ends were pointed toward the spheres. The ends of the vessels may have launched in the direction of their axis during a BLEVE and escalated the incident.

The proximity of residential population around the facility contributed to the high death toll. It is often the case that population is not near a plant when it is built, but moves around it afterwards. Thought should be given to buying extra land around a site to provide a buffer zone. Many companies now use quantitative risk techniques to estimate the risk to individuals both on and off a petrochemical site. The population density around the site is usually one of the inputs to these studies. If there is a significant change in off-site population, these studies need to be reviewed to see if conclusions from them are still valid.

3.14.5 References and Links to Investigation Reports

- Lees 1996, Lees, F.P., 'Loss Prevention in the Process Industries – Hazard Identification, Assessment and Control', Volume 3, Appendix 4, Butterworth Heinemann, ISBN 0 7506 1547 8, 1996.
- HSE Website, Case Studies (http://www.hse.gov.uk/comah/sragtech/casepemex84.htm)
- CCPS, 2008, "Incidents That Define Process Safety", American Institute of Chemical Engineers, Center for Chemical Process Safety, New York, NY, 2008.

- CCPS 2003, Guidelines for Facility Siting and Layout, American Institute of Chemical Engineers, Center for Chemical Process Safety, New York, NY, 2003.

3.15 Process Safety Culture: Methyl Isocyanate Release, Bhopal, India, 1984

3.15.1 Summary

Just after midnight on December 3, 1984, a pesticide plant in Bhopal, India released approximately 40 metric tons of methyl isocyanate (MIC) into the atmosphere. The incident was a catastrophe; the exact numbers are in dispute, however, lower range estimates suggest at least 3,000 fatalities, and injuries estimates ranging from tens to hundreds of thousands. The event occurred when water contaminated a storage tank of MIC.

3.15.2 Detailed Description

MIC is a flammable and extremely toxic liquid. It is also water reactive, with the reaction being very exothermic. There was water ingress into the MIC storage tank, and an exothermic reaction did occur. To this day, it is still not certain how the water got into the tank. There are several theories of exactly what happened. One is that water entered the tank through a common vent line from a source over a hundred meters away. Another is that water was deliberately introduced by a disgruntled employee. There are also theories that water entered from the scrubber over time because the tank was not pressurized or that there was a mix-up in the hose connections for water and nitrogen (Macleod, 2014).

Whatever the initial source of the water contamination, there were several failures of other systems that could have mitigated the consequences of the event. See Figure 3.36.

- Pressure gauges and a high temperature alarm which could have warned of the reaction failed.
- A refrigeration system that cooled the liquid MIC was shut down to save money. This could have removed heat from the reaction to prevent or reduce the amount of MIC that boiled up.
- The relief vents of the MIC tank were directed to a scrubber that could have detoxified the MIC, however, the vent gas scrubber was turned off.
- The scrubber was vented to a flare which could have burned the MIC, however, it was disconnected from the process while corroded pipework was being repaired.
- A fixed water curtain designed to absorb MIC vapors did not reach high enough to reach the gas cloud.

NEED FOR PROCESS SAFETY

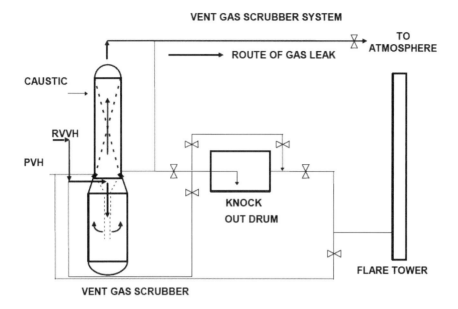

Figure 3.36. Schematic of emergency relief effluent treatment system that included a scrubber and flare tower in series, source AIChE.

3.15.3 Key Lessons

Process Safety Culture (Section 2.2). The plant in Bhopal was running under severe cost pressures because the product was selling at only 1/3 of the plant's design capacity. Not only were safety systems shutdown to save money, but maintenance of the plant itself was cut and the plant was in disrepair. There were also cuts in staffing and training. All this is evidence of a culture that prioritized cost over process safety.

Hazard Identification and Risk Analysis (Section 2.8). Inherent safety is an approach to process safety that emphasizes reducing or eliminating hazards as opposed to controlling them. Bhopal is an example of how the inherent safety strategy called **minimize** (the others are **substitute**, **moderate**, and **simplify**) could have reduced the consequences of an incident. The minimize strategy can be summed up by the saying "what you don't have, can't leak". MIC was an intermediate in the process to make the pesticide SEVIN. The Bhopal plant had three 15,000 gallon MIC tanks (Figure 3.37). Large intermediate tanks provide flexibility in a chemical process. In the case of MIC, however, a large inventory

112 INTRODUCTION TO PROCESS SAFETY FOR UNDERGRADUATES

Figure 3.37. Photograph taken shortly after the incident. A pipe rack is shown on the left and the partially buried storage tanks (three total) for MIC are located in the center of the photo right, (source Willey 2006).

meant an increased hazard. The key lesson here is to assess the hazards from the inventories of hazardous chemicals in an HIRA before deciding how much, if any, to store.

Management of Change (Section 2.14). Disabling of protective systems must be covered by a management of change review. The Bhopal plant was designed with several protection layers against an MIC release. No adjustments were made in the MIC storage tank protection strategy, such as increased monitoring or a reduction in the amount stored, as the layers were removed.

3.15.4 References and Links to Investigation Reports

- CCPS 2008, "Incidents That Define Process Safety", American Institute of Chemical Engineers, Center for Chemical Process Safety, New York, NY, 2008.
- U.S. Chemical Safety and Hazard Investigation Board, Video - Reactive Hazards - Four major incidents illustrate the dangers from uncontrolled chemical reactions (http://www.csb.gov/videos).
- Macleod 2014, Macleod, Fiona, Impressions of Bhopal, Loss Prevention Bulletin, Vol. 240, p. 3- 9, December 2014.

- Willey 2006, Willey et al., "The Accident at Bhopal: Observations 20 Years Later", Presentation to AIChE Spring National Meeting, April 2006)

3.16 Failure to Learn, BP Macondo Well Blowout, Gulf of Mexico, 2010

3.16.1 Summary

Most of the information in this section comes from a report by the Bureau of Ocean Energy Management Regulation and Enforcement (BOEMRE 2011). At approximately 9:50 PM on the evening of April 20, 2010, an undetected influx of hydrocarbons escalated to a blowout on the Deepwater Horizon rig at the Macondo Well. A cause of the blowout was failure of a cement barrier in the production casing string, a high-strength steel pipe set in a well to ensure well integrity and to allow future production. The failure of the cement barrier allowed hydrocarbons to flow up the wellbore, through the riser and onto the rig, resulting in the blowout. Shortly after the blowout, hydrocarbons that had flowed onto the rig floor through a mud-gas vent line ignited in two separate explosions. Flowing hydrocarbons fueled a fire on the rig that continued to burn until the rig sank on April 22 (Figure 3.38). Eleven men died on that evening. Over the next 87 days, almost five million barrels of oil were discharged from the Macondo well into the Gulf of Mexico (BOEMRE 2011).

3.16.2 Detailed Description

The well was being temporarily shut down. The production casing, a high strength steel pipe set up in a well to ensure well integrity and allow future production, was installed on April 18-19. It was located in a laminated sand-shale zone instead of at a consolidated shale strata.

On April 19, cementing was begun. The purpose of the cement is to seal the well and prevent hydrocarbons from flowing into the well. During the drilling of the well, there had been significant losses of drilling mud into the formation. BP engineers chose to do the cementing in way that would minimize losses. They reduced the amount of cement, the typical pumping rates and used different cement than planned. Tests run on the proposed cement mixture for its stability by Halliburton, which supplied the cement and did the cement job, were not complete at the time of the work.

After the blowout, investigations showed the cement did not meet the API RP 65 standard, *Cementing Shallow Water Flow Zones in Deep Water Wells*.

Figure 3.38. Fire on Deepwater Horizon, source (CSB, 2010).

The cement operation was monitored by comparing the amount of material flowing into the well with what comes out. The crew believed they had seen a full return of everything that went in, indicating a successful cementing job. Later examination of the data showed that up to 80 barrels (3360 gallons) of material could have been lost.

After the cementing was completed, a well integrity test was run. The results of the test showed that drill pipe pressure was increasing; this was an indication the cement barrier had failed and material was flowing into it. The test was repeated several times with negative results. Not believing the results, the crew developed a faulty theory to explain the differences. A final test, a cement bond log, which would have been a conclusive test, was cancelled on the belief the cement barrier injection was successful. The BOEMRE investigation states that the "central cause of the blowout was failure of a cement barrier in the production casing string" (BOEMRE 2011).

An extremely simplified explanation of this behavior is that, based on the original monitoring of material in and out, the crew believed the cement job was successful, and any evidence to the contrary was rationalized away. A more thorough description of this "confirmation bias" is given in Hopkins (2012.)

After deciding the cement job was successful, the crew began to complete the temporary abandonment procedures. During this time, the well was supposed to be monitored for abnormalities, specifically, a "kick" (an influx of hydrocarbon into

NEED FOR PROCESS SAFETY

the well that forces drilling mud back up into the well). Kicks are detected by imbalances in the inflow and outflow of the well. The outcoming mud is directed to a pit, and an unusual rate of change in the level in the pit indicates a kick. At this point, the crew began directing the mud to two pits instead of one, and from them to other pits and from the rig to another ship, reducing the ability to rapidly detect a kick. All of this was a violation of the rig owner's policies regarding well monitoring. When a kick did occur, it was not detected by the crew.

During this time, volume in some of the tanks and pits was increasing. The pit level rose by 100 barrels (4,190 gallons) in 15 minutes. The crew's response was to try to bleed off pressure by opening the well, an indication they still did not know that the well was actually flowing. The capacity of the mud gas separator (Figure 3.39) was overwhelmed and the hydrocarbon flowed onto the rig. The blowout could have been sent to a diverter which would have directed it off of the rig (Figure 3.40). Procedures on when to use the diverter instead of the mud gas separator were not clear, however (BOEMRE).

Figure 3.39. Location of Mud-Gas separator, source (TO, 2011).

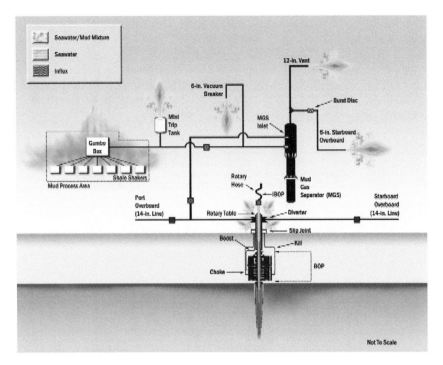

Figure 3.40. Gas release points, source (TO, 2011).

Gas alarms began going off on the rig. The general alarm system was not set to go off automatically, so after the gas alarms went off, the control room had to manually sound the general alarm. The engine room operators called for instructions, but were never told to shut down the engines. The engines were later determined to be the likely ignition source. Personnel were not told to evacuate until 12 minutes after the first gas alarm went off.

There was a Blowout Preventer (BOP), a large, 17 m (57 feet) tall, and 399 m-tons (400 tons) apparatus at the ocean floor designed to seal a well in an emergency, on the Macondo Well. It had Variable Bore Rams (VBR) designed to seal around the drill pipe and "annulars" designed to close around the drill pipe (Figure 3.41). The annulars and VBR and were activated by the crew. It also had a Blind Shear Ram (BSR), designed to cut the drill pipe and seal the well. The BSR was not activated by the crew, but was later found activated, either due to the loss of signals from the well (the design intent) or later by a remotely operated vehicle. In either case, it failed to seal the well. Later investigation showed that the drill pipe had buckled during the event and was forced outside of the zone of the blades of the BSR. (See the link for CSB Website in the Links section for a video describing the BOP operation and why it failed.)

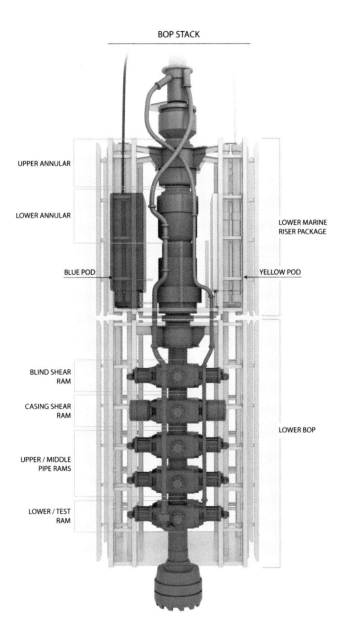

Figure 3.41. Macondo Well blowout preventer, source (CSB 2010).

3.16.3 Key Lessons

Learning from Experience. The four pillars of process safety were described in Section 2.1. The fourth pillar was *"learning from experience"*. The Deepwater Horizon well blowout was an informative illustration of the need for learning from experience.

Incident Investigation: The simplest example of not learning form experience concerns the kick that went undetected for 30 minutes. A kick had occurred previously on March 8, 2010, and it also was not detected for 30 minutes. Detection and response to a kick is a key safety barrier in well operations. The failure to detect the kick of March 8 should have been investigated. This was required by BP's own internal requirements. The failure to do an investigation was cited as a contributing cause to the incident by the BOEMRE report (p. 110).

The next level of failure to learn was not learning the lessons from similar incidents at other rigs. In 2008, a blowout occurred on a BP rig in the Caspian Sea. It was reported to be due to a poor cement job and resulted in 211 people being evacuated from the rig and the field being shut down for 4 months. In December 2009, an event similar to the Deepwater Horizon's occurred on a rig operated by Transocean in the seas off of the United Kingdom. The crew had finished displacing mud and conducted a pressure test. They stopped monitoring and were surprised when mud began flowing onto the rig. In this event they were able to shut down the well. The lessons from these events were not learned by the crew and engineers running the Deepwater Horizon. Transocean, the owner and operator of the drilling rig, prepared a presentation on this event, and issued an operations advisory to its North Sea fleet. It appears that neither the presentation nor the advisory had been sent to the Deepwater Horizon (Hopkins, 2012).

Process Safety Culture: In Section 2.2 characteristics of a good process safety culture were listed. They included **maintaining a sense of vulnerability and establishing a learning/questioning environment**. The Deepwater Horizon incident shows the consequences of a poor safety culture.

The pressure tests of the cement barrier failed, but were explained away. This is a symptom of a lack of a learning/questioning environment and a lack of a sense of vulnerability.

Further illustrating this point was the BOEMRE report statement that "in the weeks leading up to the blowout on April 20, the BP Macondo team made a series of operational decisions that reduced costs and increased risk" and that the investigation team "found no evidence that the cost-cutting and time-saving decisions were subjected to the various formal risk assessment processes that BP had in place".

3.16.4 References and Links to Investigation Reports

- Bureau of Ocean Energy Management Regulation and Enforcement (BOEMRE), Report Regarding the Causes of the April 20, 2010 Macondo Well Blowout (September 14, 2011).
  (http://docs.lib.noaa.gov/noaa_documents/DWH_IR/reports/dwhfinal.pdf)
- Hopkins, Andrew, Disastrous Decisions: The Human and Organizational Causes of the Gulf of Mexico Blowout, CCH, Sydney, AU (2012).
- CSB 2010, U.S. Chemical Safety and Hazard Investigation Board, Investigation Report No. 2010-10-I-OS, Explosion and Fire at the Macondo Well, Vol. 1 (June 5, 2014).
- Chemical Safety Board Video Room (**http://www.csb.gov/videos/**).
- TO, 2011, Macondo Well Incident, Transocean Investigation Report, Vol. 1 (June 2011).

3.17 Summary

This chapter summarized sixteen incidents showing the importance of the elements of process safety management (PSM). They are meant to raise your awareness about incident history and lessons that have been learned the hard way. Trevor Kletz, a world renowned expert in process safety, is often quoted as saying "Organizations don't have memory, only people do." By providing a group of examples, this chapter has started you off on your career of collecting the necessary memories.

The Swiss cheese model of incidents, which illustrates the idea that incidents usually have more than one cause, was also introduced. Most processes are designed with more than one protection layer. However, no protection or safeguard is 100% perfect; there are "holes" in every one. Incidents occur when multiple failures lines up, as in the holes of a piece of Swiss cheese. One of the many goals of PSM is to make a pathway through the "holes" as unlikely as possible. The layers of protection as represented as a piece of Swiss cheese are essential to prevent or mitigate a loss of containment and must be sustained by the elements of the PSM System.

A few of the incidents mentioned above represent defining moments in process industry or for the companies that experienced them and for the CPI as a whole.

The people killed in Texas City refinery explosion (Section 3.1) were not involved in the process at all, and did not need to be located so close to a hazardous unit operation. This incident led to a major revision of API code regarding facility siting: API Recommended Practice 752: Management of Hazards Associated with Location of Process Plant Permanent Buildings. It also led to the creation of API Recommended Practice 753: Management of Hazards Associated with Location of Process Plant Portable Buildings.

The Piper Alpha fire and explosion (Section 3.7) led to development of stronger offshore safety requirements in the UK Offshore Installations (Safety Case) Regulations. A Safety Case is the documentation that a production organization must submit in the UK to demonstrate that their operation is safe. Another change made was having responsibility for enforcing safety case moved from the UK's Department of Energy to the Health and Safety Executive (HSE) to avoid potential conflicts between production and safety.

The Flixborough incident (Section 3.11) is commonly cited as a prime example of the need for a good management of change program. The phrase "they didn't know what they didn't know" is frequently mentioned. It is also an example of the need for following modern facility siting codes.

An outcome of the Exxon Valdez incident (Section 3.13) was Exxon's development of its Operational Integrity Management System (OIMS). This is Exxon's safety and PSM system. Exxon's OIMS precedes the OSHA PSM regulation in the U.S. It consists of 11 elements which approximately match the elements developed by the CCPS, and are more comprehensive than the OSHA PSM standard. A more detailed explanation of the OIMS can be found at:

http://cdn.exxonmobil.com/~/media/Brochures/2009/OIMS_Framework_Brochure.pdf

As mentioned in Section 3.15, the Bhopal incident led to the formation of CCPS. It and the PEMEX explosion in Section 3.14, which occurred only a few months earlier, were key drivers in the creation the OSHA PSM regulation mentioned in Chapter 2. Union Carbide, which built the Bhopal plant, had a good reputation in the chemical industry with respect to its technical expertise. The Bhopal event shows that technical competence was not enough and that management systems also play a key role in process safety.

3.18 References

3.1 Incidents That Define Process Safety, American Institute of Chemical Engineers, Center for Chemical Process Safety, New York, NY, 2008.

3.2 Mannan, Sam, Lees' Loss Prevention in the Process Industries (Third Edition), Elsevier, Amsterdam, 2005.

3.3 Reason, James. "The Contribution of Latent Human Failures to the Breakdown of Complex Systems". Philosophical Transactions of the Royal Society of London. Series B, Biological Sciences 327 (1241): 475–484. April 12, 1990.

4

Process Safety for Engineering Disciplines

4.1 Introduction

This chapter focuses on the role of a new engineer in their first 12 to 24 months on the job in a process industry organization or plant. Chapter 2 described the 20 elements of Risk Based Process Safety Management. Table 4.1 lists the element with the activities new engineers can participate in. The roles of Chemical, Mechanical, Civil, Instrumentation and Electrical (I&E), Control and Safety Engineers and how they interface with some key process safety elements is described in this chapter.

Civil/structural/geotechnical engineers also have a role, especially in areas where seismic threats exist and where corrosion can weaken structural support members or damage secondary containment dikes, etc.

Engineering specialties such as fire protection engineering and environmental engineering are also often involved along with, of course, process safety engineering specialists.

4.2 Process Knowledge Management

When starting to work in a process plant, or doing engineering design or process development, the new engineer should learn what the process safety hazards are in the plant and process(es) he or she is assigned to, regardless of their engineering discipline. Engineers should become familiar with the existing Process Safety Information (PSI) to understand the hazards and engineering controls of the process. Chemical, Mechanical, Instrumentation and Electrical, and Control Engineers all contribute to developing and maintaining PSI.

Chemical engineers (often filling the role of the Process Engineer) are likely to be responsible for keeping information on the process, such as hazards of the chemicals, chemical reactions and reactivity hazards, heat and mass balances, relief device sizing basis and calculations, and Piping and Instrumentation Drawings (P&IDs) up to date. In research and development, and scale up, developing and organizing the PSI will be part of the chemical engineers responsibility.

Table 4.1. Process safety activities for new engineers.

Element	Activity	Discipline
Commit to Process Safety		
Process Safety Culture	Learn responsibilities for process safety roles	All
Compliance with Standards	Learn and apply standards and regulations that apply to processes and equipment	All
Process Safety Competency	Take advantage of training opportunities	All
Workforce Involvement	Contribute Listen to the input from operators, technicians, etc.	All
Stakeholder Outreach	Become aware of the organization's outreach efforts	All
Understand Hazards and Risk		
Process Knowledge Management	Ensure accuracy of PSI Be familiar with Safety Data Sheets Develop reactivity matrix Develop and update design basis calculations Develop and update Piping and Instrumentation Diagrams (P&IDs) Verify electrical classification Develop automation logic diagrams and cause and effect charts. Maintain equipment files (design and fabrication information)	All All Chemical Chemical Chemical I&E Engineers I&E & Control Mechanical
Hazard Identification and Risk Assessment	Participate in HIRAs Assemble required PSI Review applicable RAGAGEPs Compile industry incidents Civil / structural engineers role in review of seismic threats and structural response to impulse loading from overpressure events, etc.	All Chemical All All Civil Engineers
Manage Risk		
Operating Procedures	Write new operating procedures Update operating procedures	Chemical
Safe Work Practices	Write Permits Approve permits	Safety Safety & Chemical

Table 4.1. Process safety activities for new engineers, continued.

Element	Activity	Discipline
Asset Integrity and Reliability	Identify equipment covered by Inspection, Testing and Preventive Maintenance (ITPM) program	Chemical & Mechanical
	Write ITPM procedures	Mechanical
	Analyze inspection results	Mechanical
	Approve and monitor repairs	Mechanical
	Maintain ITPM records	Mechanical & I&E & Control
	Testing of SCAI and Safety Instrumented Systems	
	Review structural integrity	Civil Structural Engineers
Contractor Management	Approve contractors with respect to safety	Safety
	Train contractors	All
Training and Performance Assurance	Generally not applicable for new engineers	
Management of Change	Identify changes	All
	Participate in MOCs	All
	Assemble required PSI	All
	Review applicable RAGAGEPs	All
Operational Readiness	Participate in readiness reviews	All
	Assemble required information	All
Conduct of Operations	Perform all tasks reliably in accordance with policies and procedures	All
Emergency Management	Learn emergency response plans for your area	All
Learn from Experience		
Incidents	Participate in investigations	All
	Recognize and identify near misses	All
Metrics	Record and maintain data	All
	Analyze data	All
Auditing	Assemble requested information for audits	All
Management Review	Generally not applicable for new engineers	

Mechanical engineers (sometimes referred to in industry as Reliability engineers) will be responsible for the development of PSI regarding the equipment of the process and the management systems and activities needed to keep this information up-to-date. This can include the design codes to be followed, vessel information such as Maximum Allowable Working Pressure (MAWP), materials of construction, and piping and equipment vulnerabilities to stress cracking, thermal cycling, and stress analysis that may contribute to hazardous events. In the design phase of a project, mechanical engineers can work with chemical engineers in the equipment selection for a process by providing input on the reliability and maintainability of process equipment. More reliable equipment will increase the safety of a chemical process.

In the design phase of a project, I&E and Control engineers are involved in the design of process control systems, Safety Controls Alarms and Interlocks (SCAI) and Safety Instrumented Systems (SIS). I&E and Control engineers collaborate with chemical engineers to find out what the control system and SIS requirements are. With this information, they can specify the design of and the equipment used for process control and SIS. The list of SCAI and SIS functions is part of the PSI that an I&E or Control engineer will help to maintain. A description of the work processes involved is provided in the CCPS book *Guidelines for Safe and Reliable Instrumented Protective Systems,* 2007 (Ref. 4.1). Further instruction on inherently safer automation practices and instrumented safeguard design may be found in the CCPS book *Guidelines for Safe Automation of Chemical Processes,* 1993 (Ref. 4.2).

Safety Engineers may be responsible for the PSI concerning hazards related to chemical toxicity and personnel controls and the design basis of fire protection systems. In smaller organizations, safety engineers may have a much larger role in process safety than in a larger facility which may have a separation of occupational safety and process safety responsibilities. In these organizations, the safety engineer may have a larger role in developing and maintaining the PSI.

4.3 Compliance with Standards

All engineers should learn what regulatory obligations and standards may apply to the plant or process they work in. Examples of regulations in the U.S. include OSHA Process Safety Management and EPA Risk Management Program regulations. There can also be state and local regulations such as the California Risk Management and Prevention Program that cover facilities in certain regions. Examples of regulations in other countries include the Seveso Directive in Europe, COMAH regulations in the UK, and the Canadian Environmental Protection Act.

There are many third party standards that new engineers may have to know. It is up to each organization to decide and document which third party standards it wants to follow. New engineers should then learn what these standards are. If there is a process safety department in the organization, that would be a good source for finding out what standards and codes apply to the organization or plant.

For example, Chemical Engineers may need to know the requirements of American Petroleum Institute (API) standards and recommended practices, and NFPA fire standards and insurance guidelines. There can also be manufacturing association standards that apply to a plant or specific chemical. A small sampling to illustrate such standards is listed below.

- NFPA 30, Flammable and Combustible Liquids Code
- API 752, Management of Hazards Associated With Location of Process Plant Buildings
- FM Data Sheet 7-82N, Storage of Liquid and Solid Oxidizing Materials (https://www.fmglobal.com/fmglobalregistration)
- The Fertilizer Institute, Recommended Practices for Loading/Unloading Anhydrous Ammonia (https://www.tfi.org/safety-and-security-tools/recommended-practices-loadingunloading-anhydrous-ammonia)
- Chlorine Institute, Chlorine Customers Generic Safety and Security Checklist (http://www.chlorineinstitute.org/pub/ed19b46c-c6a7-acca-ce7f-43f496d0dcae)

Mechanical Engineers may need to know the requirements of API standards and recommended practices dealing with construction or corrosion. Also, the American Society of Mechanical Engineers (ASME) has standards dealing with topics such as pressure vessels, test codes, and piping systems. Examples of some standards a Mechanical Engineer may have to learn are:

- ASME Boiler and Pressure Vessel Code
- API RP 941, Steels for Hydrogen Service at Elevated Temperatures and Pressures in Petroleum Refineries and Petrochemical Plants
- NACE Standards

I&E and Control engineers are involved in the design of control systems, Safety Controls Alarms and Interlocks (SCAI) and Safety Instrumented Systems (SIS). Examples of standards I&E and Control engineers may need to become familiar with are:

- IEC 61508, Functional Safety of Electrical/Electronic/Programmable Electronic Safety-related Systems (E/E/PE, or E/E/PES), 2010.
- IEC 61511, Functional safety - Safety instrumented systems for the process industry sector, 2003.

- NFPA 70®: National Electrical Code®
- **ANSI/ISA-84.00.01-2004 Parts 1-3 (IEC 61511 Mod)** Functional Safety: Safety Instrumented Systems for the Process Industry Sector. ISA-84.91.xx - a series of normative standards and guidelines regarding SCAI.

In addition, the I&E and Control engineers must become familiar with the Safety Manuals of the instruments and logic solvers they use for safety systems, in order to design, install, configure and program the devices in a way that will support the risk reduction that safeguard is intended to provide.

Some third party standard organizations will allow people to view, but not download, their standards on-line at no charge. Examples of organizations that do this are the API and NFPA. FM Global, an insurance company, publishes a series of data sheets that can be downloaded at no charge. Industry groups such as the Chlorine Institute or The Fertilizer Institute also allow some of their guidelines to be downloaded at no charge.

4.4 Hazard Identification and Risk Analysis, Management Of Change

New processes and substantially modified processes often require the involvement of many engineering disciplines in addition to the role of chemical engineers. Mechanical engineers are often helpful in identifying vulnerabilities such as materials of construction, stress cracking, thermal cycling, and stress analysis that may contribute to hazardous events. Civil engineers may be needed to identify concerns/solutions to external events such as flooding, earthquakes, and high wind loading. Instrument and Control engineers are often crucial to identifying control reliability, control response, and control suitability for addressing consequences identified. Electrical engineers often provide insight into critical distribution system reliabilities, needs for redundancy, and issues involving electrical coordination.

A new engineer is very likely to participate in hazard identification reviews such as HAZOPs, What-If/Checklists, etc., and Management of Change (MOC) reviews. Participation in Process Hazard Analysis (PHA) reviews and other hazard analysis activities such as Layer of Protection Analysis (LOPA) will depend on where the process is in its PHA cycle (typically PHAs are renewed on a 3 to 5 year cycle depending on the organization and regulatory requirements). Participation in a formal HIRA such as a HAZOP is an excellent way to learn about the process to which one is assigned.

In any HIRA there is overlap and synergy in the knowledge and skills brought by the various engineering disciplines involved. Mechanical and I&E and Control engineers may or may not participate in PHAs as full time participants, although it is a prudent practice for organizations to include them in PHAs on a part-time basis. As the engineers who specify items such as equipment parameters, material of construction, and control system design, their knowledge of how a process operates, the process hazards, and how the process responds to deviations is valuable. Electrical engineers often provide insight into critical distribution system reliabilities, needs for redundancy, and issues involving electrical coordination.

Many, if not most, organizations will require the involvement of a safety engineer or representative from the safety department in incident investigations, MOC reviews, hazard reviews, pre-startup safety reviews and PHAs. Topics such as human factors, ergonomics, operator training, fire safety, and industrial hygiene are important to process safety and the design of plants and processes; however, the average chemical engineer may receive little or no training on these subjects. Therefore, during a PHA a safety engineer's input on these topics is valuable.

Almost all engineers will have some involvement in MOC reviews, depending on the need for their expertise. Everything said above about PHAs applies to MOC reviews also.

Management of Organizational Change

Over time a new engineer can expect to move into roles of increasing responsibilities, either as a supervisor/manager or technical specialist. These new roles will bring increased responsibility with regard to process safety. As stated in the introduction, some organizations have training matrices for these positions as well as new engineers.

More organizations are coming to realize that the PSM Element of Management of Change should apply to changes in positions and organizational change. Some examples of incidents in which organizational change played a role are listed in Table 4.2.

To this end, some organizations have developed MOC programs for personnel and organizational changes (sometimes referred to as Management of Organizational Change (MOOC)) as well as changes to production processes. The CCPS book *Guidelines for Managing Process Safety Risks During Organizational Change* 2013 (Ref. 4.3) covers this subject.

The trigger for an MOOC review is when a change can have an impact on process safety as well as personal safety and the environment. Just as a new engineer can become involved in an MOC review, he or she can also become involved in an MOOC review for an organizational change. These reviews are meant to identify the potential process safety impact of the change and what additional training is needed to prepare a person for the new responsibilities.

Table 4.2. Incidents with organizational change involvement

Incident	Organizational Change
BP Refinery Explosion, Section 3.1	Staffing level reduction: After staff reductions, the remaining ISOM operators were likely fatigued from working 12-hour shifts for 29 or more consecutive days.
Esso Longford Explosion, Section 3.5.	Increase in responsibility: All of the plant's process engineers were relocated to the head office in Melbourne, Australia. Supervisors and operators were given greater responsibility for operating the plant, including troubleshooting, for which they were not properly prepared.
Formosa Plastics VCM Explosion, Section 3.10	Staffing level reduction: Formosa eliminated an operator group leader position and shifted their responsibility to the shift supervisors, who were not always as available as the group leaders used to be.
Flixborough Explosion, Section 3.11	Temporary replacement: Maintenance manager role temporarily filled by a lab head pending reorganization.

Examples could be a change in staffing level or a temporary assignment (for themselves or others).

4.5 Asset Integrity and Reliability

An important, if not most important, contribution of mechanical engineers to process safety is to the Asset Integrity and Reliability element (see section 2.11). Mechanical engineers are responsible for developing test procedures, overseeing inspection programs and analyzing the data from inspection programs. To this end, a new mechanical engineer should become familiar with applicable codes and standards, as mentioned in the Compliance with Standards section above.

Mechanical engineers often have the lead role in identifying failure rates in determining reliability. This activity is particularly important to organizations that are developing in-house values in lieu of generic values derived from various publications. Failure rates will be translated into frequencies of failure and probability of failure on demand which are key values used in semi-quantitative and quantitative risk analysis.

The following CCPS books cover the topic of Maintenance and Mechanical Integrity

- *Guidelines for Mechanical Integrity Systems*, 2006.
- *Guidelines for Improving Plant Reliability through Data Collection and Analysis*, 1998.
- *Guidelines for Safe Operations and Maintenance*, 1995.

I&E and Control engineers will often be tasked with designing SCAI and safety instrumented (SIS) systems and ensuring that the required reliability and probability of failure on demand is achieved. They may often be asked to develop procedures for safety instrument system test protocols, instrument calibrations and testing, control loop response capabilities etc. Similar to the mechanical engineers role in asset reliability, instrumentation failure rates etc. may need to be identified to support semi quantitative and quantitative risk analysis.

4.6 Safe Work Practices

Safe work permits such as hot work and confined space entry are important. Safety engineers have a lead role in implementing work permit procedures, and sometimes in improving them.

While safety engineers may take the lead in developing safe work practices, other engineering disciplines also have a role. Chemical engineers are often called upon to look at what adjustments to process, what isolation, and what preparation is needed for safe work practices to be implemented in a process area. They will often be looked upon to develop lockout plans, confined space entry plans, rescue plans, isolation schemes, and equipment and process preparation activities associated with maintenance, repairs, and inspection.

Often mechanical engineers will be looking to avoid the inclusion of confined spaces or to include consideration of confined spaces in their designs. This may involve things such as providing removable spools for isolation of equipment, and additional support for piping and equipment disconnected as part of the isolation. Mechanical engineering may also be involved with limiting the flammability or combustibility of materials, identifying equipment locations that allow good access, tie off points, ease of entry, etc.

Electrical engineers often have a role in identifying lockout devices, energy isolation and coordination, the provision of try steps, and lockout design for power sources such as lighting, control panels, and power panels.

Instrument and Control engineers may need to ensure that control devices can be isolated and secure, that there is limited access to PLCs and other logic solvers, that there is adequate control of bypassing, and proper labelling and drawings to ensure safe maintenance repair activities. Also in their scope they may be developing and following validation protocols for new or modified software or programming changes for control systems. They may also be called upon to ensure the proper electrical classification has been identified and that instrumentation is appropriate for the classification and fit for use in the process.

4.7 Incident Investigation

All engineers in a plant or process must learn what the definition of process safety incidents and near misses are for your organization to be able to spot near misses. Safety engineers will frequently take the lead in incident investigations.

New engineers from many disciplines will likely get a chance to participate in process safety near miss and incident reviews, especially for more significant incidents. Chemical engineers contribute the knowledge of the process technology, and process chemistry, chemical interactions and kinetics may have contributed to the incident. Mechanical engineers contribute with the knowledge of what equipment failed, what the failure modes and causes are. Instrument and Control engineers will contribute their knowledge of how control and SCAI systems work and what new controls may be needed. Electrical engineers may identify electrical component failures, substation/MCC electrical coordination issues, contributions electrical noise and harmonics, as well as solutions and action items to address electrical reliability.

4.8 Resources for Further Learning

There are many process safety courses offered by consulting firms specializing in process safety and by the AIChE Academy. Courses can be found on topics such as:

- Process Hazard Analysis
- Management of Change
- Auditing
- Incident Investigation
- Emergency Relief System Design
- Chemical Reactivity Hazards
- Dust Explosion Hazards
- Consequence Analysis
- Quantitative Risk Analysis
- Safety Instrumented Systems and IEC61511

There are several online AIChE webinars and eLearning courses in process safety. Some are free to undergraduate and graduate student AIChE members. AIChE members get credits to apply to courses each year that covers the cost of the some of the courses. Listed below is an example of some AIChE process safety webinars or eLearning courses:

- Basics of Lab Safety
- Chemical Process Safety in the Process Industries
- Dust Explosion Control

- Essentials of Chemical Engineering for non-Chemical Engineers
- Inherently Safer Design
- Layer of Protection Analysis Process Safety 101
- Process Safety Management for Bioethanol
- Runaway Reactions
- What Every Young Engineer Should Know about Process Safety
- SACHE modules (see Chapter 6)

Non-chemical engineers working in the process industries will benefit from the *Essentials of Chemical Engineering for non-Chemical Engineers,* an online in classroom AIChE course. This course will help non-chemical engineers to learn some fundamentals of chemical engineering and enable better communication/collaboration with chemical engineers on projects.

All new engineers should sign up to receive the CCPS Process Safety Beacon (PSB) and news releases from the Chemical Safety Board (CSB) at the sites below.

- http://www.aiche.org/ccps/resources/process-safety-beacon
- http://www.csb.gov/news/

The CCPS Process Safety Beacon is a resource aimed at delivering process safety messages to plant operators and other manufacturing personnel. The monthly one-page Process Safety Beacon covers the breadth of process safety issues. Each issue presents a real-life accident, and describes the lessons learned and practical means to prevent a similar accident in your plant.

The CSB investigates and issues reports on significant process safety incidents in the U.S. Access to the reports and videos about the incidents is available through their website. By signing up for news releases, an engineer can keep up to date on what reports and videos have been issued and what investigations are ongoing. During their first year, a new engineer in the process industries should find time to look at the backlog of CSB videos. Most of the videos are about specific incidents; however a few are worth noting for a new engineer.

Any engineer working in a process plant or refinery should watch the videos:

- *Hot Work: Hidden Hazards, Dangers of Hot Work*
- *No Escape: Dangers of Confined Spaces*
- *CSB video on Valero Refinery Asphyxiation Incident*

These address hazards common to all processing facilities.

The videos *Reactive Hazards* and *Combustible Dust: An Insidious Hazard* covers specific hazards that may be encountered in certain chemical process facilities. Reactivity hazards and combustible dusts, except agricultural dusts, are not covered by regulations such as OSHA PSM. If hired into a facility where chemical reactions are run or combustible dust handled, a new engineer should see these videos. The CSB videos can be accessed at www.csb.gov/videos.

After seeing the *Reactive Hazards* video mentioned above, new engineers working in a chemical, biochemical or petrochemical processing facility should obtain a copy of the *Chemical Reactivity Worksheet*. This free software program has a database of several thousand chemicals that contains hazard and reactivity information about each chemical. The program can predict potentially harmful interactions between two chemicals and display them in a compatibility chart. The program can be found at: http://response.restoration.noaa.gov/reactivityworksheet .

Another program a new engineer should consider getting is CAMEO Chemicals. It provides physical and hazard properties of many chemicals and allows the user to create a collection of chemicals and determine the hazards of uncontrolled mixing. This program is available at:

http://cameochemicals.noaa.gov/

4.8 Summary

In the chemical process and associated industries, several engineering disciplines have a role in process safety. The role of Chemical, Mechanical, I&E, Control, and Safety engineers in some key process safety management elements, namely, Compliance with Standards, Process Safety Knowledge, Hazard Identification and Risk Assessment, Management of Change, Asset Integrity and Reliability, Safe Work Practices, and Incident Investigation has been highlighted.

Resources for further learning about process safety have been identified in the respective sections.

4.9 References

4.1 Guidelines for Safe and Reliable Instrumented Protective Systems, American Institute of Chemical Engineers, Center for Chemical Process Safety, New York, NY, 2007.

4.2 Guidelines for Safe Automation of Chemical Processes, American Institute of Chemical Engineers, Center for Chemical Process Safety, New York, NY, 1993.

4.3 Guidelines for Managing Process Safety Risks During Organizational Change, American Institute of Chemical Engineers, Center for Chemical Process Safety, New York, NY, 2013.

5

Process Safety in Design

5.1 Process Safety Design Strategies

Inherently safer design (ISD). Inherently safer design is defined as a way of thinking about the design of chemical processes and plants that focuses on the elimination or reduction of hazards, rather than on their management and control. Often, the traditional approach to managing chemical process safety accepts the existence and magnitude of hazards in a process, and works to add sufficient safeguards to reach the desired level of risk by reducing the likelihood or consequences of process safety events.

Where feasible, ISD has the potential to make the chemical processing technology simpler and more economical in many cases, and enables more robust and reliable risk management. The four strategies for designing inherently safer processes are:

- **Minimize** – use small quantities of hazardous materials, reduce the size of equipment operating under hazardous conditions such as high temperature or pressure
- **Substitute** – use less hazardous materials, chemistry, and processes
- **Moderate** – reduce hazards by dilution, refrigeration, process alternatives which operate at less hazardous conditions
- **Simplify** – eliminate unnecessary complexity, design "user friendly" plants

Resources for ISD include *Inherently Safer Chemical Processes – A Life Cycle Approach* (Ref. 5.1) and *Process Plants - A Handbook for Inherently Safer Design* (Ref. 5.2).

The order of preference for general process safety design strategies is:

Inherent. Eliminate or greatly reduce the hazard by changing the process or materials to use materials and conditions which are nonhazardous or much less hazardous.

Passive. Passive strategies minimize hazards using process or equipment design features which reduce the frequency or consequence of an incident without the active functioning of any device. A containment dike is an example of a passive safeguard. A passive safeguard can degrade with time, for example, a concrete dike can develop cracks and holes, so they must be inspected accordingly.

Active. Active strategies include process control systems, safety interlocks, automatic shutdown systems, and automatic incident mitigation systems such as sprinkler systems to extinguish a fire. Several elements have to function for active systems to work. These systems must therefore be tested in a regular basis.

Procedural. Procedural safety features include standard operating procedures, safety rules and procedures, operator training, emergency response procedures, and management systems.

When implementing the safety design strategies, the goal should be to use controls that prevent the event first, such as safety interlocks, sand then controls that mitigate the effect of the event, such as fire suppression systems.

5.2 General Unit Operations and Their Failure Modes

The following sections describe the process hazards of common unit operations and types of equipment found in chemical, biochemical, petrochemical and other types of facilities. The second section of this chapter will deal with operations common to petroleum oil refineries.

5.2.1 Pumps, Compressors, Fans

Overview. Pumps, compressors and fans are used to move fluids from one point to another. In doing so, they impart energy, in the form of pressure and temperature, to the fluid being moved. If they are run with the inlet and or outlet blocked they can heat the contained fluid, which can have consequences depending on the characteristics of the fluid. This can create hazards that will depend on the properties of the fluid being moved. As rotating equipment items, they will have seals around rotating shafts, whose failure can lead to leaks. Again, the hazards from leaks will depend on the fluid being moved. Finally, they can just fail to run or run for too long, leading to potential hazardous consequences to other parts of the process.

Common failure modes for pumps and compressors include stopping, deadheading and isolation, cavitation/surging, reverse flow, seal leaks, casing failures, and motor failures.

Knowledge of the properties of the fluid is necessary to assess the hazards of the potential failures of fluid transfer equipment. Deadheading or isolation of pumps and compressors can lead to uncontrolled reactions, exothermic decomposition, and explosion hazards when moving chemicals that are reactive, have thermal stability issues, or are shock sensitive because pumps and compressors impart energy to the fluid. An example of this would be pumping Ammonium Nitrate (AN) solutions described in Section 3.6, which become more sensitive to deflagration and detonation at high temperatures. If loss of

PROCESS SAFETY IN DESIGN 135

containment due to a seal failure occurs, then the release of materials that are flammable can lead to fire and explosion hazards:, also corrosive or toxic materials can create personnel hazards. A low boiling fluid can flash, so knowledge of the vapor pressure/temperature is needed.

The material presented here is an overview of design considerations. See Section 6.6 of *Guidelines for Engineering Design for Process Safety*, 2nd Edition (Ref. 5.3) for more details.

Example Incidents.

Example 1. The pump in Figure 5.1 was destroyed because the mechanical seal failed. The light hydrocarbon being pumped was released; it ignited and burned – causing extensive local damage. No one was near the pump when the fire occurred, so there were no injuries (Ref. 5.4).

Example 2. A 75 HP centrifugal pump was operated with both suction and discharge valves closed for about 45 minutes. It was believed to be completely full of liquid. As mechanical energy from the motor was transferred to heat, the liquid in the pump slowly increased in temperature and pressure until finally the pump failed catastrophically (see Figure 5.2). One fragment weighing 2.2 Kg (5 lbs.) was found over 120 meters (400 feet) away. Luckily, no one was in the area so there were no injuries (Ref. 5.5).

Figure 5.1. Damage from fire caused by mechanical seal failure.

Figure 5.2. Pump explosion from running isolated.

Design considerations for process safety. There are different types of pumps and compressors, for example, centrifugal and positive displacement. When pumps and other rotating equipment are running, the process fluid can leak from between the rotating shaft and the body of the pump. Leaks can result in fires or toxic releases if the fluids are flammable or toxic. There are different types of seal configurations to prevent these leaks. The selection of pump and seal type is usually dependent on process considerations. There are process safety implications for every type of pump and seal.

With compressors, liquid entry into the compressor can cause catastrophic failure. Protection should be provided upstream of compressors to remove liquids and associated shutdown should systems should also be provided.

Centrifugal pumps. Centrifugal pumps (Figure 5.3) are susceptible to leaks, deadheading, running isolated, cavitation and reverse flow. Design configurations that have two pumps in parallel can be especially vulnerable to these failure modes because the possibility of starting the wrong pump is always present.

Centrifugal pumps, as with other rotating equipment, need shaft seals between the process fluid and the external environment. The simplest form of a seal is a packing material. This can degrade with time and leak. Mechanical seals are the next type. In a mechanical seal pump, a seal face is kept in contact between the shaft and casing. These seals leak less than packing, but do require a lubricating

PROCESS SAFETY IN DESIGN

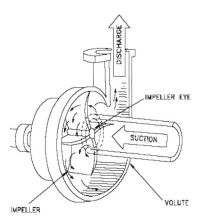

Figure 5.3. Schematic of centrifugal pump, Ref. 5.6.

fluid that must be compatible with the process fluid. There are single and double mechanical seals (Figure 5.4). In a double mechanical seal there are two seals back to back inside a chamber external to the pump. The seal chamber is flushed with a fluid, and leaks are contained and can be detected in this fluid.

There are also sealless pumps. Pumps with a magnetic drive, where there is no direct connection between the motor and the pump shaft, are an example of these. If a very hazardous fluid is being pumped, sealless pumps can be the best choice.

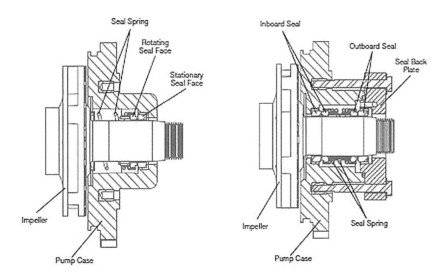

Figure 5.4. Single and Double Mechanical Seals, Ref. 5.7.

Sealless pumps also have safety considerations. If they are run dry (i.e. the inlet is blocked), the bearings can be damaged, causing high temperature. There is a risk tradeoff in the selection of centrifugal sealed pumps in comparison to sealless pumps. Pumps with seals may fail more frequently with lower consequence, while sealless pumps may experience catastrophic failures, but less frequent, failures. For example, a sealless pump may be appropriate for a highly toxic fluid but not for a less hazardous one. As with all risk decisions, a company's risk acceptance criteria must be consulted during selection.

The hazards of running deadheaded or isolated were described in the first two examples above. In a deadheaded pump, a blockage on the discharge side of the pump results in the flow reducing to zero and an increase in the discharge pressure. The energy input from the deadheaded pump increases the temperature and pressure of the fluid in the pump. Designs should be considered to operate in a manner that prevents the pump from a deadhead operation for more than a very short period of time.

If a centrifugal pump stops while on line, the fluid can flow in reverse, from the destination to the source if the piping and differential pressures allow it. Hazard Identification and Risk Analysis (HIRA) (Section 2.8) is needed to assess the consequences and protections needed for the reverse flow scenario.

Positive displacement pumps. There are numerous types of positive displacement (PD) pumps, such as the rotary screw pumps, gear pumps, as shown in Figures 5.5 – 5.6 and diaphragm pumps. Also included in this category are progressive cavity, piston pumps and peristaltic tubing pumps, some with large capacity. Reverse flow is more difficult in positive displacement pumps, but positive displacement pumps can build up high pressures if deadheaded. Air driven diaphragm pumps can, however, be operated deadheaded. Because of the rapid buildup of pressure if many types of positive displacement pumps are deadheaded, some type of pressure relief or automatic shutoff device triggered by a pressure sensor is almost always included with the installation of positive displacement pumps. Many companies will also install a pressure relief device or high pressure shutoff external to the pump.

Process Safety and Specifications. Table 5.1 lists the common failure modes, consequences and design considerations for pumps, compressors, fans. No single table or source can list all the potential consequences of a given failure mode, this table is provided as a starting point only.

PROCESS SAFETY IN DESIGN 139

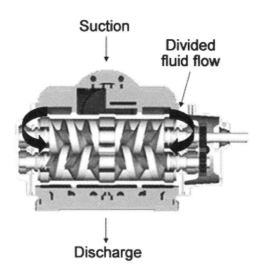

Figure 5.5. Two-screw type PD Pump, courtesy Colfax Fluid Handling.

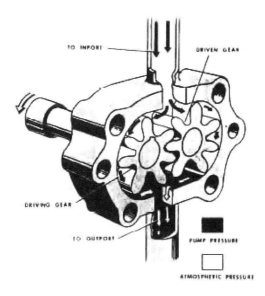

Figure 5.6. Rotary Gear PD pump, source http://www.tpub.com/gunners/99.htm.

Table 5.1 Common failure modes, causes, consequences, design considerations for fluid transfer equipment

Failure mode	Causes	Consequences	Design considerations
Stopping	Power failure Mechanical failure Control system action (failure or intended)	Consequence to upstream or downstream equipment (HIRA needed) See Reverse Flow	Power indication on pump Low flow alarms/interlocks Level alarms and interlocks in other equipment
Deadheading or Isolation	Pump/compressor outlet blocked in by: Closed valves (manual, control, block) on discharge side, Plugged lines Blinds left in	Loss of containment due to, high temperature and pressure causing seal, gasket, expansion joint, pump or piping failure. Possible phase changes, reactions.	Overpressure protection. Minimum flow recirculation lines. Alarms/interlocks to shut down the pump or compressor on low flow or power Limit closing time for valves
Cavitation / Surging	Blocked suction by: Closed inlet valves Plugged filters/strainers	Loss of containment due to damage to seals or impellers,	Low flow alarms/interlock to shut down the pump or compressor Vibration alarms/interlocks
Reverse Flow	Pump or compressor stops	Loss of containment upstream Overpressure upstream Contamination upstream	Non-Return (Check) valves on discharge side[1] Automatic isolation valves Overpressure protection upstream Positive displacement pump
Seal Leaks	Particulates in feed Loss of seal fluids or flushes Small bore connections Age (wearing out)	Loss of containment due to damage to seals	Alarms or interlocks on seal fluid system to shutdown pump/compressor Double mechanical seals with alarm on loss of one seal Sealless pumps
Contamination / change of fluid	Liquid in compressor feed	Compressor damage See Seal leaks	Knock out pots before compressor

[1] Check valves are tricky to count on, their dangerous failure modes are difficult to diagnose or test for until they are actually needed.

Figure 5.7 shows a Pump Application Data Sheet. The first block of information, Liquid Properties, specifically asks for safety information such as flammability, toxicity, regulatory coverage. Other properties that could be of interest could be thermal stability or reactivity of the fluid. The next block, Materials of Construction, is important to safe processing. Use of the incorrect material of construction can lead to loss of containment.

Engineering standards that include design considerations for pumps are:

- API STD 610, Centrifugal Pumps for Petroleum, Petrochemical and Natural Gas Industries, Eleventh Edition (ISO 13709:2009 Identical Adoption), Includes Errata (July 2011)
- API STD 685, Sealless Centrifugal Pumps for Petroleum, Petrochemical, and Gas Industry Process Service, Second Edition, Feb. 2011
- API STD 674, Positive Displacement Pumps-Controlled Volume for Petroleum, Chemical, and Gas Industry Services, 3rd Edition, Includes Errata (June 2014)
- API STD 617, Axial and Centrifugal Compressors and Expander-compressors, September 2014

5.2.2 Heat Exchange Equipment

Overview. Heat exchange equipment is used to control temperature by transferring heat from one fluid to another. Heat transfer equipment includes heat exchangers, vaporizers, reboilers, process heat recovery boilers, condensers, coolers and chillers. Much of what is stated in this section will also apply to heating/cooling coils in a vessel such as a reactor or storage tank.

Failures in heat transfer equipment can lead to loss of temperature control, contamination of one of the fluids or loss of containment. Temperature is frequently a critical process variable, so failure of this equipment due to fouling, plugging, or loss of the heat transfer fluid supply can lead to serious consequences. An HIRA is needed to assess these. Due to its nature, heat exchange equipment can see thermal stress due to temperature gradients. This can lead to loss of containment. The Longford fire and explosion in Section 3.5 is an example of this failure mode.

Leaks due to corrosion or erosion are another common failure mode. The consequence of this depends on the nature of the process, the direction of the leak (process side to utility or vice versa), and the fluids involved. Each combination

Pump Application Data Sheet

OEC FLUID HANDLING INC.

Equipment No.: _____
NAME _____
COMPANY _____
ADDRESS _____
CITY _____ STATE _____ ZIP _____

Date: _____
TITLE _____
PHONE _____
FAX _____
EMAIL _____

Send completed worksheets to:
OEC Fluid Handling Inc.
P. O. Box 2807
Spartanburg, SC 29304
Fax: 1-864-573-9299
Email: sales@oecfh.com

LIQUID PROPERTIES
Pump #: _____
Liquid: _____
Pump Temp °F: _____
SP. Gravity @ P.T.: _____
Viscosity: _____ ☐ SSU ☐ CPS ☐ Other: _____
PH: _____ % Solids: _____

Safety/Environmental:
☐ Flammable ☐ Explosive ☐ Carcinogenic ☐ Toxic
☐ Noxious ☐ Regulated ☐ FDA ☐ EHEDG
Comments: _____

SYSTEM
Discharge Pressure Required (PSIG): _____
Capacity (US GPM) Max: _____ Min: _____
Suction Lift: _____
Suction Conditions: _____

Duty Cycle: ☐ 24/7 ☐ 8-10 hrs ☐ Intermittent
Comments: _____

MOTOR/DRIVER REQUIREMENTS
☐ Electric Motor ☐ Engine ☐ Air ☐ Other: _____

Enclosure:
☐ ODP
☐ TEFC
☐ TENV
☐ EX. Proof
☐ Encap
☐ Inverter Duty
☐ Mag Drive
☐ DC Drive

Voltages:
☐ 3-60-230/460
☐ 3-60-208
☐ 3-50/60-208-220/440
☐ 3-60-575
☐ 1-60-115/230
☐ 3-50-200/400
☐ 3-50-220/380
☐ 3-50-115/230
☐ 3-50-220/440
☐ 3-50-550
☐ Specify Voltages not listed above: _____
Phase: _____ Cycles: _____ Volts: _____

Additional Data: UL Label, fugitive emissions; tropical windings, motor heaters, special enclosures, etc.
(Specify): _____

Special Drives: ☐ V-Belt ☐ Inverter ☐ Air Motor
☐ Special (Specify): _____
Comments: _____

MATERIALS OF CONSTRUCTION
☐ Cast Iron ☐ CPVC
☐ Ductile Iron ☐ Hastalloy
☐ 316 Stainless Steel ☐ Alloy 20
☐ PVDF ☐ Other: _____

Casing Connections: ☐ NPT ☐ Flanged ☐ Other: _____
Jacketing for cooling/heating: ☐ Yes ☐ No
O'ring Material: ☐ Buna ☐ TFE ☐ Viton ☐ Other: _____

STUFFING BOX
☐ Mechanical Seal ☐ Packing
Preferred Seal Mfg.: ☐ Graphite
☐ Cartridge ☐ Double ☐ Other: _____
☐ Single ☐ Lip
☐ Other: _____
Make: _____ Type: _____
Material: _____ Gland Type: _____
Comments: _____

BASE PLATE/MOUNTING
Pump Mounted: ☐ Horizontally ☐ Vertically
Base Plate: ☐ Fab Steel ☐ Chan Steel ☐ Cast Iron
Coupling: ☐ Jaw ☐ Spacer ☐ Other _____
Alignment Lugs: ☐ Yes ☐ No
Comments: _____
Painting: ☐ None ☐ Mfg. Std. ☐ Primed Only
Special Painting (Specify): _____

CUSTOMER REQUIREMENTS
Drawings: _____
Approval Dimensional Drawings: ☐ Yes ☐ No
Testing: _____
Hydro: ☐ None ☐ Witness ☐ Non-Witness
Performance: ☐ None ☐ Witness ☐ Non-Witness
Field: ☐ None ☐ Witness ☐ Non-Witness
Inspection prior to shipping: ☐ Yes ☐ No
Start-up Assistance: ☐ Yes ☐ No
Operator Training: ☐ Yes ☐ No
Maintenance Training: ☐ Yes ☐ No

Serving the Fluid Handling Process Needs of Industry | 1-800-500-9311 | 1-864-573-9200 | www.oecfh.com

Figure 5.7. Example application data sheet, courtesy of OEC Fluid Handling.

PROCESS SAFETY IN DESIGN 143

has its own unique process safety issues. Failure to keep the fluids separate due to tube leaks can result in reactive chemical incidents (see example 1), or release of a toxic or flammable material into the low pressure side where it can escape elsewhere, such as at a cooling water tower.

Example Incidents

Example 1: A plant had an explosion in the outlet piping of an oxidation reactor which ruptured a 36 inch (0.9 m) pipe (see Figure 5.8). The explosion was caused by the reaction of molten nitrate salt, used to remove heat from the reactor, leaking into the piping where carbonaceous deposits had been trapped in a short dead-leg. Reactive chemical testing indicated that the reaction resembled closely the decomposition of TNT explosive. Fortunately, nobody was injured. The incident showed that it was critical to avoid leaks of the nitrate salt, to detect leaks if they did occur and to have a safe shutdown procedure if there was a leak (Ref. 5.8).

Figure 5.8. Ruptured pipe from reaction with heat transfer fluid.

Example 2. Section 3.5 described the explosion in the Longford gas plant in Australia. Process upsets led to the lean oil pumps being tripped off (stopped by an interlock). The loss of lean oil flow caused temperature drops in heat exchangers, eventually resulting in cracks in a heat exchanger when hot oil was reintroduced to the cold section of the heat exchanger. This led to a gas release and explosion. An HIRA is needed to detect consequences such as these.

Design Considerations. There are several different types of heat exchangers: shell and tube, plate and frame, spiral, and air cooled are examples. See Figures 5.9 – 5.11.

One of the biggest concerns is mixing of fluids due to tube leaks. Design considerations to prevent or mitigate this include:

- Put a highly toxic fluid on the tube side so tube leaks go into the shell side and can be detected in the cooling tower or piping at low non-hazardous concentrations. Also leaks from a shell failure are the utility fluid and not the highly toxic fluid.
- Careful selection of materials of construction to resist corrosion on both sides.
- Design for drainage to reduce corrosion by installing exchanger in a sloped orientation (consider the baffle design to allow fluids to drain).
- Use of double tube sheets for heat exchangers handling toxic chemicals or for materials where mixing must be avoided (see Figure 5.12).
- Consideration of fluid velocities, fluid properties, contaminants (solids and dissolved materials), and impingement.

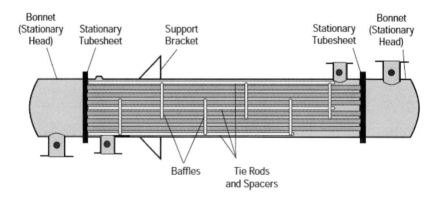

Figure 5.9. Shell and tube heat exchanger, Ref. 5.9.

PROCESS SAFETY IN DESIGN 145

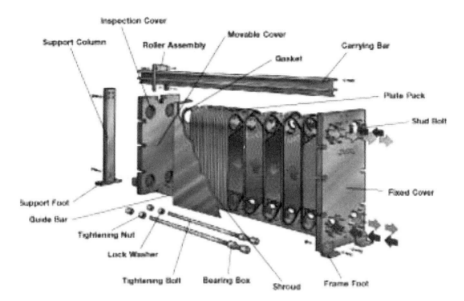

Figure 5.10. Cutaway drawing of a Plate-and-Frame Heat Exchanger, Ref. 5.10

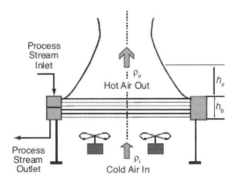

Figure 5.10. Schematic of air cooled heat exchanger, Ref. 5.11.

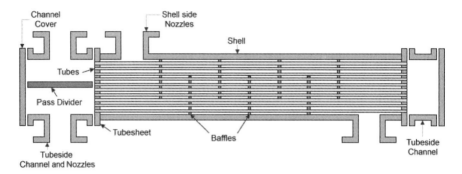

Figure 5.12. Double tube sheet, courtesy www.wermac.org

Control of temperature in a process is usually an important, if not critical, process parameter. Some design considerations to deal with this are:

- Design for periodic cleaning to remove fouling.
- Provide a tube sheet vent nozzle or other a means to vent non-condensable gases from the process system.
- Tube pitch and spacing, flow distribution, fluid velocity and ΔT should be considered to prevent fouling.

Engineering standards that include design considerations for heat exchangers are:
- ASME Boiler and Pressure Vessel Code (Ref. 5.3)
- API Standard 660, Shell and Tube Heat Exchangers (Ref. 5.10)
- Tubular Exchanger Manufacturers Association (TEMA)
- Heat Exchanger Institute standards

Table 5.2 presents some common failure modes and design considerations for heat exchangers.

5.2.3 Mass Transfer; Distillation, Leaching and Extraction, Absorption

Overview. Mass transfer operations are used to separate materials, purify products, and detoxify waste streams. Knowledge of the properties of the materials being handled is necessary to assess the hazards of the potential failures of mass transfer equipment.

Distillation (see Figure 5.13), *stripping* and *absorption* frequently involve flammable materials; therefore, loss of containment can result in fires and explosions. High temperatures are used, especially in the reboilers, to drive the distillation/stripping; therefore the thermal stability of the materials being handled

Table 5.2 Common failure modes, causes, consequences, design considerations for heat exchange equipment.

Failure Mode	Causes	Consequences	Design Considerations
Leak from heat transfer surface	Corrosion from contaminants in the process fluids, and cooling fluids, and/or loss of treatment chemicals. Anaerobic attack under sediments and scale. Thermal stress (e.g. extreme heat/cold)	Loss of containment Inadvertent mixing and contamination of low pressure side, potential reactions, (HIRA needed)	Periodic inspection Choice of materials of construction Choice of heat transfer fluid Shell expansion joints Non shell and tube design Control of introduction of process fluids during startup and shutdown Monitoring of low pressure side fluid Toxic fluids in tubes, monitor shell side. Treatment chemicals
Rupture from heat transfer surface	Corrosion Thermal stress (e.g. extreme heat/cold) Operation out of design temperature range resulting in stress cracking, improvement, weakening of tubes or tubesheet (see loss of cooling or heating load) Blocking in one fluid side during operation	Potential rupture of heat exchanger Loss of containment	Emergency relief device Control of introduction of process fluids during startup and shutdown

Table 5.2 Common failure modes, causes, consequences, design considerations for heat exchange equipment, continued.

Failure Mode	Causes	Consequences	Design Considerations
Loss of cooling or heating fluid	Loss from supply Control system malfunction Pluggage Mis-valving	Loss of process control (HIRA needed) High pressure	Alarms / interlocks on low flow or pressure of heat transfer medium High or low temperature alarms on process side
Inadequate heat transfer	Fouling Accumulation of non-condensable gases (mostly condensers)	Loss of process control (HIRA needed) High pressure	Ability to clean High or low temperature alarms on process side

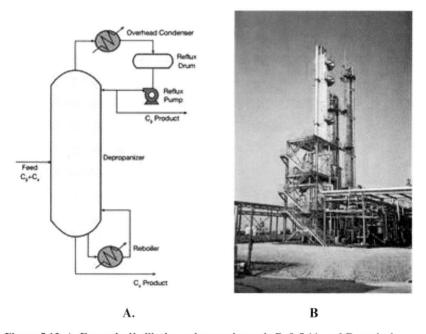

A. B

Figure 5.13. A. Example distillation column schematic Ref. 5.11, and B. typical industrial distillation column, ©Sulzer Chemtech Ltd.

needs to be understood. Loss of cooling to a reflux condenser can affect the composition of materials in a distillation, which again leads to the need to understand the effect of composition on the thermal stability characteristics of the material being handled. The Concept Sciences explosion in Section 3.4 is an example of a failure that led to a higher, and more dangerous, concentration than expected than intended. High levels of liquid in columns can lead to plugging of internals, high pressure, and loss of containment. The Texas City explosion in Section 3.1 is an example of this. Higher liquid loading on trays can result to damage to trays and result in more serious temperature upsets.

Packing material fires. Hydrocarbon residue that remains on column packing can self-ignite at elevated temperatures when exposed to the atmosphere. Iron sulfide, which is pyrophoric, can form from sulfur found in crude oil. Corrosion of carbon steel components can settle on packing and can ignite when exposed to the air or oxygen (Ref. 5.14 and 5.15).

Adsorption. Adsorption processes are exothermic. Carbon bed adsorbers are subject to fires due to this overheating. For certain classes of chemicals (e.g. organic sulfur compounds (mercaptans), ketones, aldehydes, and some organic acids) reaction or adsorption on the carbon surface is accompanied by release of a heat that may cause hot spots in the carbon bed. Adsorption of high vapor concentrations of organic compounds also can create hot spots. If a flammable mixture of fuel and oxygen are present, the heat released by adsorption or reaction on the surface of the carbon may pose a fire hazard (e.g., a fire may start if the temperature reaches the autoignition temperature of the vapor and oxygen is present to support ignition) (Ref 5.16 and 5.17). Figure 5.14 is a schematic of a carbon bed system, In Figure 5.14, the top bed is in absorption mode and the bottom bed is in recovery mode.

Extractors. Extractors will contain two immiscible fluids plus some materials being transferred from one phase to another. Loss of containment can result in flammable or toxic releases, depending on the nature of the materials. Failure of level control in extractors can result in the wrong material being sent to downstream equipment, leading to high levels or pressure in downstream equipment.

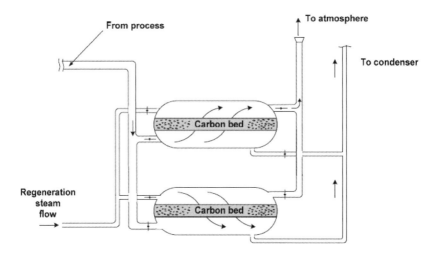

Figure 5.14. Schematic of carbon bed adsorber system, Ref. 5.16.

Example Incidents.

Distillation Column Incident. In 1969, an explosion occurred in a butadiene recovery unit in Texas City, Texas. The location of the center of the explosion was found to be the lower tray section of the butadiene refining (final purification) column. The butadiene unit recovered byproduct butadiene from a crude C4 stream. The overhead product of the refining column was a high-purity butadiene product. The heavy components of the feed stream, including vinyl acetylene (VA), were removed as a bottoms product. The bottoms vinyl acetylene concentration was normally maintained at about 35%. Explosibility tests had indicated that VA concentrations as high as 50% were stable at operating conditions. Highly concentrated VA, on the other hand, decomposes rapidly on exposure to high temperature.

When the butadiene unit was shut down to undertake necessary repairs, the refining column was placed on total reflux. The refining column explosion occurred approximately 9 hours after it was placed on total reflux. This operation had been performed many times in the past without incident. The operators did not observe anything unusual about this particular switch over to total reflux. Subsequent examination of the records indicated that the column had been slowly losing material through a closed, but leaking, valve in the column overhead line.

Loss of butadiene through the leaking valve resulted in substantial changes in tray composition in the lower section of the column. The concentration of vinyl acetylene in the tray liquid in the vicinity of the tenth tray apparently doubled to an estimated 60%. The loss of liquid level in the base of the column uncovered the reboiler tubes, allowing the tube wall temperature to approach the temperature of the steam supply. The combination of increased vinyl acetylene concentration and high tube wall temperature led to the decomposition of VA and set the stage for the explosion that followed (Ref. 5.18, 5.19 and 5.20).

Carbon Bed Incident. A bio-solids thickener tank in a refinery was overpressured by an internal explosion. The tank roof traveled about 200 feet. The tank was vented to a 55-gallon activated carbon adsorber to remove organics and control odors. After two years of operation, seal covers had been installed on the tank as part of an emissions control program. On the day of the event, fresh carbon was installed in the drum. The heat of adsorption increased the temperature in the drum to about 350 °F. The hydrocarbons adsorbed on the carbon oxidized and a local hot spot was formed. Over several hours, the temperature increased to the point where the carbon started burning at about 800 °F. This ignited the inlet vapors to the adsorber and flames propagated back to the tank, igniting the vapor space.

Prior to the tank being sealed, hydrocarbons were able to escape to the atmosphere. When the tank was sealed better, this created a higher organic load to the carbon unit which generated more heat and provided fuel for the flames to propagate back to the tank (Ref. 5.21).

Design Considerations. Distillation is temperature, pressure, and composition dependent; special care must be taken to fully understand any potential thermal decomposition hazards of the chemicals involved.

Columns need adequate instrumentation for monitoring and controlling pressure, temperature, level and composition. The location of sensing elements in relation to column internals must be considered so that they provide accurate and timely information and are in direct contact with the process streams.

A design feature of some columns is to provide a tall base (e.g., 3 m (10 ft.)) to provide adequate Net Positive Suction Head (NPSH) to ensure that the bottoms pumps do not cavitate and fail.

Column support structures and skirts should be fireproofed, as they are not cooled by internal fluid flow and a ground fire can lead to the column collapsing.

Overpressurization can result from freezing, plugging, or flooding of condensers, or blocked vapor outlets, if the heat input to the system is not stopped.

Emphasis should be placed upon the use of inherently safer design alternatives using concepts such as:

- Limiting the maximum heating medium temperature to safe levels
- Selecting solvents which do not require removal prior to the next process step
- Using a heat transfer medium that prevents freezing in the condenser

- Locating the vessel temperature probe on the bottom head to ensure accurate measurement of temperatures, even at a low liquid level
- Minimizing column internal inventory
- Avoiding dead legs that can corrode, plug or freeze

To prevent packing fires:

- Cool columns to ambient temperature before opening
- Wash the column thoroughly to remove residues and deposits
- Use chemical neutralization to remove pyrophoric material
- Purge columns with nitrogen
- Monitor temperatures of the packing and column as it is opened
- Minimize the number of open manways to reduce air circulation

To protect against carbon bed fires:

- Test the impact of the vapors on the carbon for potential heat release before putting the carbon adsorption system into service; if possible identify reactions that are not already known.
- Measure the off-gas for carbon monoxide and CO_2 to warn of hot spots
- Measure bed temperatures at a large number of places
- Provide fire control systems such as water sprays, nitrogen or steam
- Include flame arrestors to prevent the spread of fire from the carbon containers to the flammable chemical containers

5.2.4 Mechanical Separation / Solid-Fluid Separation

Overview. Mechanical separators are used to separate solids from liquids or gases. Typical equipment includes:

- Centrifuges
- Filters
- Dust collectors

Common failure modes for centrifuges include mechanical friction from bearings, vibration, leaking seals, static electricity, and overspeed. Vibration is both a cause of problems and an effect from other sources. Static charges can occur from the flow of the slurry and liquor into the unit, and the high speed of centrifuges. Static charges can accumulate due to the use of synthetic, non-conductive filter media. Both mechanical friction and static can ignite flammable liquids if used.

A concern for filters is exposure or loss of containment during opening and closing. Plate and frame filters have a high potential for leaks and should be avoided when flammable or toxic solvents are being filtered. The filter media can be overpressured during the end of a cycle, which can cause further leaks.

Studies have shown that dust collectors are the equipment item most frequently involved in dust explosions (Ref. 5.22). Frequently, dust collectors are at the end of a process and collect the smallest particle size dusts, which are going to be the most hazardous if the dust is combustible. Common failure modes for dust collectors include loss of containment due to failure of the filter media (usually bags or cartridges), plugging of the filter media, and loss of grounding of filter bags. The filter media is often purged with air to prevent plugging. Failure of the purge system followed by reactivation can create a much worse dust cloud. Dust collectors with filter bags (often called baghouses) can have over a hundred bags, each one is a point where the bag cage can be isolated from ground (a metal bag cage, if electrically isolated, can develop a significant electrical charge). See the example case history.

Example Incidents

Batch centrifuge. A crystalline finished product was spinning in a batch centrifuge when an explosion occurred. The product had been cooled to 19°F (-7°C) before it was separated from a methanol-isopropanol mixture in the centrifuge. It was subsequently washed with isopropanol precooled to 16°F (-9°C). The mixture was spinning for about 5 minutes when the explosion occurred in the centrifuge. The lid of the centrifuge was blown off by the force of the explosion. The overpressure shattered nearby glass pipelines and windows inside the process area (up to 20 meters away), but nearby plants were not damaged. No nitrogen inerting was used and sufficient air was drawn into the centrifuge to create a flammable atmosphere. Sufficient heat could also have been generated by friction to raise the temperature of the precooled solvent medium above its flash point. Because the Teflon® coating on the centrifuge basket had been worn away, ignition of the flammable mixture could also have been due to metal-to-metal contact between the basket and the bottom outlet chute of the centrifuge, leading to a friction spark. A static discharge might also have been responsible for the ignition. Since the incident, the company has required use of nitrogen inerting when centrifuging flammable liquids at all temperatures (Ref. 5.23).

Lessons learned include monitoring the oxygen concentration in conjunction with inerting and sealing the bottom outlet to minimize air entry. Because the ignition source was uncertain (static discharge, frictional heat), this incident illustrates why it often is prudent to assume an ignition source when designing for flammable materials. In his book, *Lessons from Disaster: How Organizations have No Memory and Accidents Recur*, Trevor Kletz is quoted as saying "Ignition source is always free." (Ref. 5.24).

Dust Collector Explosion: An explosion occurred in a dust collector connected to the process vents for polyester plastic extrusion equipment. The explosion was safely vented. The bags were off their cages and charred, as shown in Figure 5.15. There were 144 cages in the dust collector. During the investigation it was found that one of them was not grounded. A check of an identical unit revealed problems with the grounding between the cages and the tube sheet, as shown in Figure 5.16. It was also found that the type of bag used had been changed, but the need for a grounding strap was not understood, and that grounding/bonding checks were not part of the mechanical integrity program.

The type of bag used was changed to ones with an improved grounding strap to ensure grounding of the metal cage support inside the bag, and the procedures were fixed to include conductivity checks (Ref. 5.25).

Figure 5.15. Damage to dust collector bags, Ref. 5.25

PROCESS SAFETY IN DESIGN 155

Figure 5.16. Tube sheet of dust collector, Ref. 5.25.

Design Considerations. If a centrifuge or filter is used for flammable or toxic materials, fully enclosed units with automatic filtration, washing and discharge cycles should be considered to avoid exposure or loss of containment. Inerting or use of an inert gas sweep or blanket should be strongly considered if flammable liquids are being used. Pressure or vacuum filters can be used in place of centrifuges to reduce the problems due to bearing failure or vibration. Some typical centrifuges and filters are shown in Figures 5.17 and 5.18.

When designing dust collectors, it is essential to know whether the dust involved is combustible. Most organic and metal dusts less than about 500 microns in diameter are not only combustible, but under certain circumstances can, if ignited, lead to flash fires and/or explosions. The explosion severity and minimum ignition energy of the dust should be measured. Tests should be done on the most representative materials possible. Commonly used protection measures for dust collectors are explosion vents, explosion suppression systems, and inerting. Figure 5.19 shows a schematic of a typical baghouse. Figure 5.20 shows a picture of a vented explosion from a dust collector.

156   INTRODUCTION TO PROCESS SAFETY FOR UNDERGRADUATES

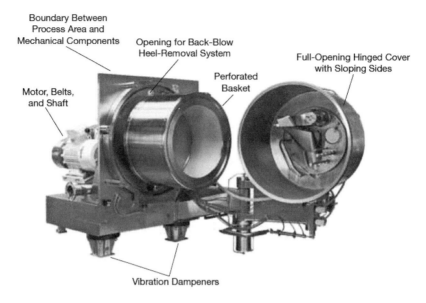

Figure 5.17. A horizontal peeler centrifuge with a Clean-In-Place system and a discharge chute, (Ref. 5.26).

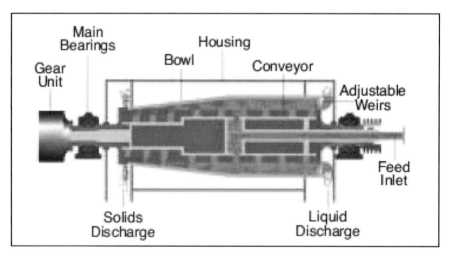

Figure 5.18. Cross sectional view of a continuous pusher centrifuge (Ref 5.26).

PROCESS SAFETY IN DESIGN 157

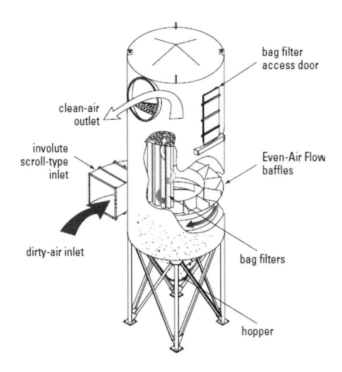

Figure 5.19. Schematic of baghouse, courtesy Donaldson-Torit.

Figure 5.20. Dust collector explosion venting, courtesy Fike.

5.2.5 Reactors and Reactive Hazards

Overview. The key process safety concern in the design of reactors is *runaway reactions*. Runaway reactions occur when the heat generation rate from an exothermic reaction exceeds the rate at which heat can be removed, causing an uncontrolled rise in temperature. If the heat released by the reaction exceeds the cooling capacity, the reaction rate will accelerate (runaway) and may result in an excessive gas evolution or a vapor pressure increase that, in the absence of adequate overpressure relief protection, can rupture the reactor. If overpressure relief protection is adequate, then there will be loss of containment through the relief device.

During runaway reactions, the temperature can rise significantly, which may favor additional exothermic reactions. If this occurs, the composition may shift to produce a more toxic off-gas, as occurred in Seveso, Italy (see Section 2.1 and the example incident below). If there is the potential for a runaway reaction, the characteristics and composition of off-gases should be understood. Appropriate downstream systems to capture and hazardous materials should be provided.

Common failure modes for reactors include: agitation failure, cooling system failure, mischarges (too much or too little of a reactant charged), or the wrong reactant charged, or reactants charged in the wrong order, and reactant quality (wrong concentration, reactant beyond shelf life).

During the reaction, the reactants and solvents need to be well mixed for the reaction to proceed as planned and for efficient input or removal of heat. Thus loss of agitation can be a cause of a runaway reaction. See the description of the Seveso incident below. A subset of agitator failure is starting an agitator too late. This allows a buildup of reactants that then suddenly are brought into contact with each other. Loss of cooling or insufficient cooling can likewise be a cause of a runaway reaction, as in the T2 Industries incident detailed below.

Examples of mischarges leading to runaway reactions would be an undercharge of a solvent meant to absorb some of the heat of reaction, or overcharging a material that could result in a more exothermic reaction than the system was designed for. Charging a reactant that is at a higher concentration than expected is an example of this. Many reactions involve a catalyst. Using a catalyst that is past its recommended shelf life or undercharge of a catalyst can lead to the buildup of unreacted material that can then react and liberate more heat than the reactor was designed for.

Another category of reactive hazards is when reactions occur where you don't want them due to inadvertent mixing. The methyl isocyanate release in Bhopal (Section 3.15) is an example of this.

PROCESS SAFETY IN DESIGN

Example Incidents

Seveso, Italy (July 10, 1976). A batch reactor was used to make 2,4,5-trichlorphenol (TCP) in two stages. Stage 1 was the reaction of 1,2,4,5-tetrachlorobenzene, and sodium hydroxide at 170-180 °C in the solvents ethylene glycol and xylene. Normally, at the end of the reaction, half of the ethylene glycol was removed by distillation and the batch cooled to 40-50 °C . Steam, normally at 190 °C, was used to heat the batch and for the distillation step. It was known that a runaway reaction could occur at 230 °C.

A batch was started on a Friday, but the plant had to be shut down for the weekend. The distillation was in progress but not completed when the reactor had to be shut down. As other parts of the plant were being shut down, the steam temperature to the reactor rose to 300 °C. At about 5 AM Saturday the reactor was shut down, and the agitator was shut off, but the reactor was not cooled down. The walls of the reactor were at 300 °C. About 7.5 hours later, a runaway reaction occurred, bursting the rupture disk and releasing about 2 Kg of dioxin, a very toxic reaction by-product, into the atmosphere. The dioxin reached nearby residential areas. Many people developed Chloracne, a skin disease. A 17 km^2 (6.6 mile^2) area was made uninhabitable, killing thousands of farm animals and contaminating soil.

The residual heat in the upper section of the reactor raised the temperature of the upper section of the liquid to about 200-220 °C (Figure 5.21). Investigations after the event showed that a slow exotherm begins at 185 °C, which could cause a

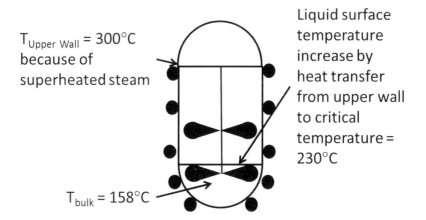

Figure 5.21. Seveso Reactor, adapted from SACHE presentation by Ron Willey.

57 °C adiabatic temperature rise, and another exotherm could start at 225 °C causing a 114 °C temperature rise. Therefore the residual heat was more than enough to raise the batch temperature to above the exotherm onsets (Ref. 5.27).

T2 Laboratories: "On December 19, 2007, a powerful explosion and subsequent chemical fire killed four employees, injured 28 members of the public who were working in surrounding businesses and destroyed T2 Laboratories, Inc., a chemical manufacturer in Jacksonville, Florida. Debris from the reactor was found up to one mile away, and the explosion damaged buildings within one quarter mile of the facility. See Figures 5.22 through 5.24.

T2 was producing a batch of methylcyclopentadienyl manganese tricarbonyl (MCMT). At 1:23 PM, the process operator had an outside operator call the owners to report a cooling problem and request they return to the site. Upon their return, one of the two owners went to the control room to assist. A few minutes later, at 1:33 PM, the reactor burst and its contents exploded, killing the owner and process operator who were in the control room and two outside operators who were exiting the reactor area." (Ref. 5.28).

The CSB found that a runaway exothermic reaction occurred during the first step of the MCMT process. A loss of sufficient cooling during the process likely resulted in the runaway reaction, leading to an uncontrollable pressure and temperature rise in the reactor. The pressure burst the reactor; the reactor's contents ignited, creating an explosion equivalent to 1,400 pounds of TNT.

Figure 5.22. T2 Laboratories site before and after the explosion, Ref. 5.28.

PROCESS SAFETY IN DESIGN 161

Figure 5.23. T2 Laboratories blast, Ref. 5.28.

Figure 5.24. Portion of 3 inch thick reactor, Ref. 5.28.

The CSB identified the following root cause:

T2 did not recognize the runaway reaction hazard associated with the MCMT it was producing.

The CSB identified the following contributing causes:

1. The cooling system employed by T2 was susceptible to single-point failures due to a lack of design redundancy.

2. The MCMT reactor relief system was incapable of relieving the pressure from a runaway reaction.

A video about the T2 Laboratories explosion can be found on the CSB website at http://www.csb.gov/videos/.

Design Considerations. The surface area to volume ratio drops as reactor size increases. The volume, and hence mass, in a reactor increases with the cube of the diameter: however the surface area, through which heat transfer occurs, increases with the square of the diameter. Therefore as a reactor gets bigger, the amount of potential heat released increases faster than the ability to remove that heat. Strategies to cope with this include adding coils for heat transfer (see Figure 2.2) or recirculating the reactants through an external cooler.

The least desirable way to run an exothermic reaction is in a *batch* reactor where all the reactants are added at one time and the reaction started. This is how the T2 Laboratories process described earlier was run. A better way to run an exothermic reaction is in a *semi batch* mode. In this mode, one or more reactants are added to the reactor gradually over the batch cycle. This enables the operator to stop the feeds if there is indication of anything going wrong, such as loss of cooling or agitation. The best way to run an exothermic reaction is in a continuous reactor (see Figures 2.1 and 2.2). For the same production capacity, a continuous reactor will be smaller, and hence have better heat removal capabilities. In some processes, reactions are run in tubular reactors, essentially a heat exchanger.

Reactors will almost always need an Emergency Relief System (ERS) to relieve the pressure from a runaway reaction. The Design Institute for Emergency Relief Studies (DIERS) has done research on the reactive relief design. This is a complex topic and a Subject Matter Expert (SME) is usually needed to do the actual ERS sizing. A hazard assessment is needed to determine appropriate relief design scenarios and chemical reactivity testing is needed to design these systems.

PROCESS SAFETY IN DESIGN 163

Alternatives or supplements to an ERS are adding chemicals such as inhibitors to stop the reaction (known as shortstopping), or rapidly emptying the reactor contents to dump tanks with water and chemicals to stop the reaction.

Inadvertent mixing can be prevented by using dedicated charging lines and through operator training and written procedures.

Table 5.3 lists some common failure modes and design considerations for reactors. See Appendix C for a Checklist for Inherently Safer Chemical Reaction Process Design and Operation.

5.2.6 Fired Equipment

Overview. Fired equipment, such as flares, incinerators, thermal oxidizers, or heat transfer fluid heaters, are commonly used to provide heat to processes, and dispose of combustible waste streams from processes. For example natural gas is converted to hydrogen in a reformer, which is a series of catalyst-packed tubes heated to several hundred degrees centigrade by a burner. Other uses of fired equipment include steam generation, and heating of distillation column reboilers.

A common failure mode is accumulation of unburnt fuel due to loss of flame, too much fuel being fed, or insufficient air (oxygen) as examples. The unburnt fuel can then ignite and cause a fire or explosion. The second most common failure mode is tube failure, which can be caused by, overheating, flame impingement, improper firing, thermal cycling, thermal shock, or corrosion. This can also result in fires and explosions. Liquid carry over into flares and incinerators can also cause explosions (see Texaco Milford Haven Explosion, Section 3.9).

Example Incidents.

Example 1. A heater was heavily damaged during startup as a result of a firebox explosion (Figure 5.25). The operator had some difficulty with the instrumentation and decided to complete the start up by bypassing the safety interlocks. This allowed the fuel line to be commissioned with the pilots out. The main gas valve was opened and gas filled the heater. The heater exploded destroying the casing and several tubes. Fortunately, no one was injured (Ref. 5.29).

Example 2. An explosion destroyed the furnace and adjacent column at a NOVA Chemical Bayport, TX plant (Figure 5.26). Before the explosion, an operator noticed flame stability problems with the low NO_x burners and began to manually adjust the airflow. During the few minutes that adjustments were being made to manage the burners, a loud puff was heard followed by a major explosion in the

Table 5.3 Common failure modes, causes, consequences, design considerations for reactors.

Failure mode	Causes	Consequences	Design considerations
Loss of Cooling	Loss of heat transfer medium from supply Control system failure	Potential runaway reaction	Emergency relief system Dual cooling modes, e.g. overhead condenser and reactor jacket Automatic actuation of secondary cooling medium on detection of low coolant flow, or high pressure, or high reactor temperature Automatic stopping of feeds of reactants or catalyst (with semi-batch or continuous reactors)
Loss of agitation	Loss of power Motor failure Agitator blades become loose/fall off	Potential runaway reaction	Emergency relief system Uninterrupted power supply backup to motor Agitator power consumption or rotation indication interlocked to stop the feed of reactants or catalyst or activate emergency cooling
Overcharge of reactant or catalyst	Error in measurement Control system failure	Potential runaway reaction Overflow of reactor	Emergency relief system Dedicated charge tanks sized to hold only the amount of reactant/catalyst needed Quantity of reactant/catalyst added limited by flow totalizer Redundant flow totalizers High level interlock / permissive to limit quantity of reactant/catalyst
Wrong reactant / catalyst	Misidentification Mix-up during product change	Potential runaway reaction	Emergency relief system Dedicated feed tank and reactor train for production of one product Control software preventing charge valve or pump operation until correct material bar code has been scanned
Step done out of sequence	Poor instructions/training Human Error	Potential runaway reaction	Controllers that verify a step has been done before advancing to the next step

PROCESS SAFETY IN DESIGN 165

Figure 5.25. Damaged heater, Example 1.

Figure 5.26. Heater and adjacent column at NOVA Bayport plant, Example 2.

furnace. It appears that the explosion was caused by clogging in the nozzles on the burners resulting in an unstable flame (Ref. 5.30).

Example 3. After a shutdown for maintenance, a hydrogen reformer in an ammonia plant was being restarted. In the normal start-up procedure at the plant, nitrogen gas is passed through the primary reformer and a heating rate of 50 °C per hour is maintained at reformer outlet. This nitrogen flows in a closed loop, that is, it is recycled back into the reformer. This cycle continues until the temperature of 350 °C is obtained at the reformer outlet. To increase reformer outlet temperature, more burners are ignited.

Because of an emergency shutdown, sufficient nitrogen inventory was not available at site for startup. At least 8 to 10 more hours were required for nitrogen inventory makeup. To save production loss, the startup procedure was initiated. Furnace firing was started in the absence of nitrogen gas, and reformer outlet temperatures were monitored for a 50 °C per hour heating rate. Reformer outlet temperatures were not increasing, so the firing rate was increased. During this period, many alarms appeared on the control system for convection zone temperatures. The alarms were inhibited to avoid any inconvenience to the control panel operator, because he was busy with the steam drum level control. As there were no changes in these outlet temperatures, the firing rate was further increased, and 56 of 72 burners were fired. This represents about 70% of the heat input, without any fluid flow through the reformer. The board operator instructed the plant operator to have a physical check of the reformer. The operator found that the reformer tubes were melting down inside the furnace.

The furnace was being fired and reformer outlet temperatures were being monitored without introduction of any nitrogen through the reformer. Because of the absence of any flow through the reformer, its outlet temperature did not increase and the increase in heat with no process flows resulted in high-tube temperatures and finally melting of the tubes (Ref. 5.31).

Design Considerations. Process controls and process safety controls are handled by two control systems. Process safety considerations are usually handled by the Burner Management Systems (BMS). The BMS monitors temperatures, pressures, and the burner flames. Interlock trips and permissive interlocks are part of the BMS controls that include the ignition sequence, fuel shutoff, and purging, i.e., making sure excess fuel is purged before relighting occurs. A BMS is a critical safety system; it should either never be bypassed, or, if an organization believes it necessary, bypassed only after a management of change review where safety controls to be used in its place are established.

PROCESS SAFETY IN DESIGN   167

Combustion Control Systems control fuel to air ratios, firing rates, etc. accepting operator input, and adjusting to system demands. Combustion control systems may include process interlocks as well.

Tube rupture may be prevented by monitoring the tube skin temperatures (preferred) or monitoring the flow through the tubes.

In boilers, loss of the boiler water level supply could be catastrophic. Reliable level monitoring and control is paramount. Reliable level and control includes the design of a continuous supply of boiler feed water.

The hazards of fired equipment are so well known that many countries have specific industry fired equipment standards which define the design features and management systems required of a BMS. Standards that cover fired equipment include:

- NFPA 85: Boiler and Combustion Systems Hazards Code, 2011.
- NFPA 86: Standard for Ovens and Furnaces, 2011
- NFPA RP 87: Recommended Practice for Fluid Heaters, 2015.
- API RP 556: Instrumentation, Control, and Protective Systems for Gas Fired Heaters, Second Edition, American Petroleum Institute. 2011.
- API RP 560. Fired Heaters for General Refinery Service, Fourth Edition, American Petroleum Institute. 2007.

5.2.7 Storage

Overview. Storage of raw materials, intermediates and final product is necessary in a processing plant. Storage vessels include pressurized storage tanks, atmospheric storage tanks and silos/hoppers (for solids). Knowledge of the properties of the material is necessary to assess the hazards of a storage tank.

Common failure modes for storage tanks include:

- Loss of containment due to overfilling, mechanical failure, overpressurization, vacuum failures
- Internal fires or explosions caused by static electricity
- Uncontrolled reactions caused by loading the wrong material into a tank.
- Boilover, the expulsion of contents caused by a heat wave from the surface burning at the top of the tank reaching a water stratum at the bottom of the tank.
- Rollover, the spontaneous and sudden movement of a large mass of liquid from the bottom to the top surface of a storage reservoir due to the instability caused by an adverse density gradient.

Example Incidents.

Buncefield Explosion and Fire. A delivery of petrol (gasoline) from a pipeline started to arrive into a tank in the Buncefield depot on a Sunday morning. The safety systems in place to shut off the supply of petrol to the tank to prevent overfilling failed to operate. Gasoline cascaded down the side of the tank. Up to 300 tons of gasoline escaped from the tank (Ref. 5.32).

About 45 minutes later, a series of explosions took place. The main explosion was massive and appears to have been centered on car parking lots just west of the depot. This explosion was a Deflagration to Detonation Transition (DDT)[2] event. This was probably aided by the dense vegetation in the area which created sufficient confinement to cause the DDT. The occurrence of a detonation, which produces much higher overpressures than a deflagration, was a surprise to experts who, prior to the event, did not expect a gasoline tank farm vapor cloud explosion could make such a transition.

These explosions caused a huge fire which engulfed more than 20 large storage tanks over a large part of the Buncefield depot. The fire burned for five days, destroying most of the depot (Figures 5.27 and 5.28). In addition to destroying large parts of the depot, there was widespread damage to surrounding property and disruption to local communities. Some houses closest to the depot were destroyed and others suffered severe structural damage. Other buildings in the area, as far as 5 miles (8 km) from the depot, suffered lesser damage, such as broken windows, and damaged walls and ceilings.

Tank Collapse. In 1919 a 2.3 million gallon (8,700 cubic meters) tank of molasses suddenly broke apart, releasing its contents into the city of Boston. A wave of molasses over 15 feet (5 m) high and 1600 feet (50 m) wide surged through the streets at an estimated speed of 35 mph (60 kph) for more than 2 city blocks (Figure 5.29). 21 people were killed and over 150 injured. The tank was not properly inspected during construction and not hydrotested before filling it. Leaks between the welds had been observed, but no action was taken (Ref. 5.33).

Wrong Chemical Unloaded. A delivery truck arrived at a plant with a solution of nickel nitrate and phosphoric acid named "Chemfos 700" by the supplier. A plant employee directed the truck driver to the unloading location, and sent a pipefitter to help unload. The pipefitter opened a panel containing 6 pipe connections, each

---

[2] The transition phenomenon resulting from the acceleration of a deflagration flame to detonation via flame-generated turbulent flow and compressive heating effects. At the instant of transition a volume of pre-compressed, turbulent gas ahead of the flame front detonates at unusually high velocity and overpressure.

PROCESS SAFETY IN DESIGN 169

Figure 5.27. Buncefield before the explosion and fires, Ref. 5.32.

Figure 5.28. Buncefield after the explosion and fires, Ref 5.32.

Figure 5.29. Molasses tank failure; before and after.

of which fed to a different storage tank. Each unloading connection was labeled with the plant's name for the material stored in the tank. The driver told the pipefitter he was delivering Chemfos 700.

Unfortunately, the pipefitter connected the truck unloading hose to the pipe adjacent to the Chemfos 700 pipe, labeled "Chemfos Liq. Add." (Figure 5.30). This is similar to the human factor issue in the Formosa Plastics explosion (Section 3.10). The "Chemfos Liq. Add." tank contained a solution of sodium nitrite. Sodium nitrite reacts with Chemfos 700 to produce nitric oxide and nitrogen dioxide, both toxic gases. Minutes after unloading began, an orange cloud was seen near the storage tank (Figure 5.35). Unloading was stopped immediately, but gas continued to be released. 2,400 people were evacuated, and 600 residents were told to shelter in place (Ref. 5.34).

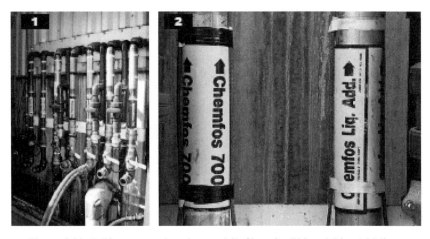

Figure 5.30. 1) Pipe connections in panel 2) Chemfos 700 and Liq. Add lines.

PROCESS SAFETY IN DESIGN   171

Figure 5.31. Cloud of nitric oxide and nitrogen dioxide.

Vacuum Collapse.: During painting, a tank's vacuum relief valve was covered with plastic to prevent potential contamination of the contents. When liquid was pumped out the covering prevented air/nitrogen from replacing the liquid volume. A vacuum developed, which led to the partial collapse of the tank, as shown in Figure 5.32 (Ref. 5.35).

Design Considerations. There are several options available to the designer when designing a storage tank. Tanks can be located underground or above ground. Above ground tanks can be fixed roof or floating roof. Finally, storage tanks can be atmospheric or pressurized.

The advantage of Underground Storage Tanks (UST) is that they cannot be exposed to an external fire from, for example, loss of containment of a flammable material from a tank in the same dike. USTs are also sheltered from swings in external temperatures. Underground tanks, however, have an increased risk of soil and/or groundwater contamination from leaks. Most underground tanks are now required to be double walled tanks or in a vault, with leak detection in the space between the tank walls or in the vault (Figure 5.37). Because of the risk of soil and groundwater contamination, the US EPA and many states have strict regulations

Figure 5.32. Tank collapsed by vacuum.

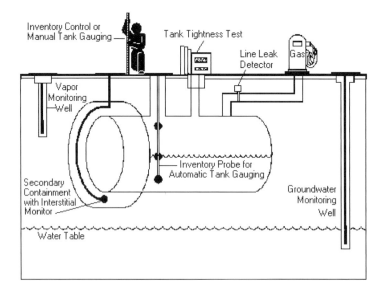

Figure 5.33. Schematic diagram of UST leak detection methods, courtesy EPA, Ref. 5.36.

covering them. The US EPA has a website with information about underground storage tanks at http://www.epa.gov/oust/index.htm

A variant of the underground tank is the mounded tank design (Figure 5.34). This is an earth covered aboveground tank. The earth cover makes the tank almost immune to BLEVE. Earth covered tanks are frequently used for an LPG bullet tank. There is no groundwater contamination issue with LPG, and an impervious membrane can be installed during construction.

Fixed roof storage tanks are usually used for materials with a vapor pressure below 1.5 psia at some specified temperature such as 20 °C. Floating roof tanks can be used with materials with a vapor pressure up to 11.5 psia, at some specified temperature. Floating roof tanks can be open or have a secondary structural roof (Figure 5.35a. and 5.35b.). The advantage of a floating roof tank is that there is no vapor space above it; therefore, if the stored material is flammable, there is no flammable vapor in the headspace that can ignite.

Floating roofs introduce other failure modes, however. The floating roof can tilt and become wedged in one position. In that case filling or emptying the tank could lead to materials getting above the roof, which can result in loss of containment from tank collapse. Also, there can be leaks between the tank wall and the floating roof seal, which, if flammable, can cause annular fires at the wall. The roof drain, the crooked pipe in the middle of the tank in Figure 5.35A, is a jointed pipe which is intended to drain rainwater into the tank dike. If it leaks, the

Figure 5.34. Mounded underground tank, courtesy BNH Gas Tanks.

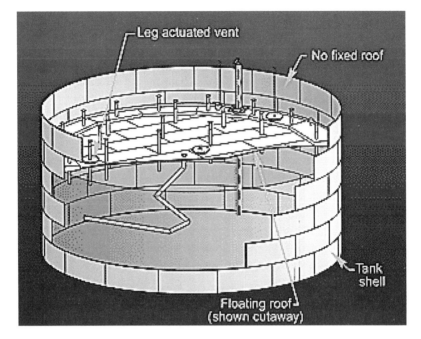

a.

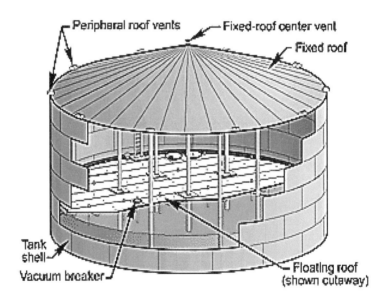

b.

Figure 5.35. Schematics of external (a) and internal floating (b) roof tanks, courtesy of petroplaza.com.

entire contents of the tank can be released if there is no detection and response mechanism.

The failure of the roof drain highlights another aspect of above ground tanks; they must be inside dikes that provide secondary containment for leaks from the tank. The dikes need to be large enough to contain the tank volume plus some safety factor (an industry rule of thumb is 110% of the volume of the tank). Local and federal regulations usually require the dike to contain the entire contents of the largest tank within a dike. The dike needs to be maintained to prevent leaks.

Pressurized storage tanks are used for materials with higher vapor pressures, such as ammonia, butane or Liquefied Petroleum Gas (LPG) (Figure 5.36). Pressurized storage tanks are susceptible to a phenomenon called boiling liquid expanding vapor explosion (BLEVE). A BLEVE occurs when a vessel containing liquid above its normal boiling point and under pressure fails catastrophically. Although there is generally a design factor of safety of up to 4 for pressure vessels (i.e., a vessel designed for 100 psi would not be expected to fail until 400 psi), the vessel may fail below its design pressure if the vapor space is exposed to the flames. This is because the ultimate tensile strength of the metal reduces to 50% of its original strength at 550 °C. Hydrocarbon flames are around 1150 °C. Therefore, the tank can fail catastrophically when the vapor space, an unwetted portion of the tank, reaches this temperature. Upon failure, the hot liquid flashes, generating a large amount of vapor and a pressure blast. If the vapor is flammable it will typically ignite and create a large fireball. The most common cause of a BLEVE is exposure to external fire. The flames heat and weaken the metal, causing it to fail. Many of the explosions in the Mexico City event in 1984 were BLEVEs (see Section 3.14).

Figure 5.36 Pressurized gas storage tank.

A fixed water spray, deluge system, or firewater monitor nozzles can keep vessels cool enough, maintaining mechanical integrity when exposed to a fire. LNG tanks are now constructed with a double wall and insulation between the outer wall and the tank to reduce heat input from a fire. Use of a mounded storage tank, mentioned earlier in this section, is also an option to protect pressurized tanks from exposure to fire.

General failure modes.

Overfilling. Overflow protection can consist of overflow lines to a safe place or instrumentation that automatically shuts off flow into the tank if the material is toxic or flammable. The overflow control system should have multiple devices to provide both redundancy and independence.

Mechanical Failure. Corrosion is a major cause of structural failure. Improper manufacture or a change in service can lead to structural failure. Age and exposure to the humid environment can cause corrosion to the point of failure over time. If the tank is insulated, corrosion under insulation (CUI) can also lead to failure. Proper choice of the material of construction, following the correct codes and standards in construction and ongoing inspection, and maintenance and testing are the main safeguards against these. New engineers may be asked to visits vendor shops to inspect and verify storage vessels meet the design specifications.

Overpressurization and vacuum. Filling or emptying a tank too quickly can cause overpressure or vacuum. Rapid cooling, for example after steam cleaning or filling with a hot material, can also cause a vacuum collapse. Pressure and vacuum protection can be installed. NFPA 30, *Flammable and Combustible Liquids Code*, (Ref. 5.37) outlines the sizing of vents and emergency vents. The operator needs to know what the design rates are for filling and emptying so as not to exceed them. High and low pressure interlocks can be used to stop filling or emptying. Frangible or weak-seam roofs can also be provided for fixed roof atmospheric tanks if needed for overpressure relief.

The overflow line for an atmospheric tank provides the overpressure protection for most atmospheric tanks and must be sized accordingly. Typically atmospheric tanks are not designed for any significant pressure beyond a few inches of liquid height above the overflow. Vent lines should be separate from overflow lines on atmospheric tanks since they are normally sized for gas flow (breathing) and are too small for liquid flow. Combining the vent and overflow lines into one large line could result in two phase flow and subsequent back pressure resulting in over pressurization of the tank.

Clogging or blocking of vents can defeat pressure/vacuum protection systems (see the vacuum collapse example). Watch out for bird nests for externally located storage tanks. Polymerizing materials can plug vents. Cold weather can result in the freezing of vents and overflow seal pots which can also result in defeating pressure/vacuum protection. Regular inspection and maintenance of pressure/vacuum protection devices is necessary.

Rollover. Rollover can result if the material can stratify in the tank. One example occurs in LNG tanks. Depending on the source of the LNG it can be of different densities. The different-density LNG can layer in unstable strata within the tank. The layers may spontaneously roll over to stabilize the liquid in the tank. Pressure relief systems may not be adequate for rollover. The force of the shifting mass can result in cracks or other structural failures in the tank. A control system of distributed temperature sensors and a pump-around mixing system can be used to provide rollover protection.

Internal fires/explosions. An internal deflagration is possible if a flammable material is being stored. Static electricity is a common form of ignition. Static can be generated by the flow of fluid through pipes, or free fall of a liquid, or by the mixing of different phases in a tank, especially if one of the phases is non-conductive. The design for flammable liquids should avoid free fall of liquid by using a dip pipe or bottom feeding. Fill rates should be kept below certain levels until a dip pipe is covered. Guidance on fill rates to minimize generation of static electricity is provided in *Avoiding Static Ignition Hazards in Chemical Operations* (Ref. 5.38). Inerting of the vapor space is another possible ignition control method. Tanks should be properly grounded to allow dissipation of static charges from all sources. Attached equipment should be bonded to the tank. NFPA 77, *Recommended Practice on Static Electricity*, (Ref. 5.39) contains information about the generation and control of static charges. The CSB video "Static Sparks Explosion in Kansas" describes an example of an explosion in a storage tank caused by static electricity. Another important safeguard needed on flammable storage tanks is a flame arrestor to prevent the flames of an external fire from propagating into the tank through the normal vent. A flame arrestor is a device that allows the gas to pass through it but stops a flame.

Lightning strikes are another common cause of ignition in storage tanks. NFPA 780, *Standard for the Installation of Lightning Protection Systems*, (Ref 5.40) provides guidance for protection of structures containing flammable liquids. A lightning strike could ignite vapors in the vicinity of the seal of a floating roof atmospheric storage tank. Usually, floating roof tanks are fitted with a foam dam around the circumference and firefighting foam chambers to add foam just to the dam and not the whole roof area. This extinguishes the fire without having to cover the whole roof. An internal floating roof tank is less susceptible because the vapor space is not flammable.

Uncontrolled reactions. Addition of incompatible materials can cause reactions, see the *Wrong Chemical Loaded* example. The first step in prevention is identification of potentially incompatible materials that could be unloaded in the storage tank. The MSDS of the material is the first place to look. Other sources include:

- *Bretherick's Handbook of Reactive Chemical Hazards, 7th Edition*, Academic Press (Ref. 5.41).
- Chemical Reactivity Worksheet, NOAA Office of Response and Restoration (Free Download) http://response.restoration.noaa.gov/oil-and-chemical-spills/chemical-spills/response-tools/downloading-chemical-reactivity-worksheet.html

If incompatible materials that could be unloaded are identified, design measures can include: positive identification of materials by sampling before unloading, locating storage tanks of incompatible materials in separate dikes, use of dedicated unloading stations with special fittings, clear labeling of unloading lines and storage tank, and clear operating procedures with written checks for material identification. If storage tanks unload into manifolds where other

PROCESS SAFETY IN DESIGN   179

materials can be, precautions against backflow into the tank include check valves, or block valves interlocked to close if backflow is detected.

Self-reacting materials, such as monomers, or water reactive materials are special cases. Temperature control, for example, cooling, may be necessary for some self-reactive materials in warm climates. Monomers are shipped with inhibitors and have a shelf life, so the tanks can be sized for rapid turnover. Monomers can also plug normal and emergency vents, so the frequency of inspection and cleaning may need to be increased. Water reactive materials can have inert gas padded atmospheres to prevent water ingress.

Codes and Standards that cover storage tanks include:

- API STD 650. *Welded Steel Tanks for Oil Storage*, 11th Edition, American Petroleum Institute. Washington, DC, 2008.
- API STD 651. Cathodic Protection for Aboveground Petroleum Storage Tanks. American Petroleum Institute. Washington, DC, 2014.
- API STD 620. *Design and Construction of Large, Welded, Low-pressure Storage Tanks*, American Petroleum Institute. Washington, DC, 2008.
- API STD 2000. *Venting Atmospheric and Low-pressure Storage Tanks*, Sixth Edition: American Petroleum Institute. Washington, DC, 2008.
- ASME Boiler and Pressure Vessel Code (Ref. 5.2).
- NFPA 30. *Flammable and Combustible Liquids Code,* National Fire Protection Association. Quincy, MA, 2008.
- NFPA 58. *Liquefied Petroleum Gas Code*, 2008 Edition, National Fire Protection Association. Quincy, MA, 2008.

5.3 Petroleum Processing

The petroleum refining sector has been selected as an example to describe within the process industries because such facilities are present in most parts of the world. They include a variety of major hazards, and, unfortunately, have a history of accidents over the last 50+ years. Furthermore, the safeguards they utilize, and modes of failure of such safeguards, are relevant to many other sectors of the process industries.

Refineries include combinations of various unit operations and process units that convert crude oil into light end products (LPG), fuels (gasoline, diesel, kerosene, jet fuel), and heavier products (lube oils, asphalt, coke). The basic operations in a refinery involve separation, breakdown of large molecules (hydrotreaters and hydrocrackers), rearranging molecules (isomerization), and combining molecules (reforming, alkylation) to turn crude oil into components such as propane, gasoline, kerosene, diesel fuel, and so on. The processes discussed below represent a sample of the units in a refinery and are not meant to be a complete list of refinery processes. Figure 5.41 shows a refinery flow diagram. *Petroleum Refining in Nontechnical Language, 4th Edition* (Ref. 42) and *An Oil Refinery Walk-Through* (Ref. 43) are good starting points for a description of refinery processes.

5.3.1 General Process Safety Hazards in a Refinery

Refineries handle large quantities of flammable gases and liquids. Any event that causes loss of containment has the potential to lead to a fire and explosion. Refineries can have flammable gas detectors with water deluge systems to limit the impact of some flammable releases. After the explosion in Texas City, described in Section 3.1, increased attention was placed on the placement and protection of buildings within refineries.

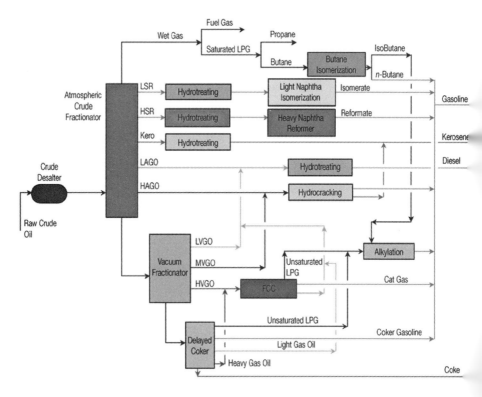

Figure 5.37. Refinery flow diagram, Ref. 43.

Many refineries handle feedstocks that contain various forms of sulfur and produce high levels of hydrogen sulfide (H_2S) as a byproduct in processing sections. H_2S is a highly toxic, dense gas, and causes death at concentrations as low as about 400 ppm. Although known for a rotten egg like odor at low concentrations, at around 100 ppm people become desensitized to its odor, so odor cannot be relied upon to provide adequate warning of exposure, or increasing concentration. In fact, 100 ppm is the Immediately Dangerous to Life and Health (IDLH) level of H_2S. This means that starting at 100 ppm H_2S can pose an immediate threat to life causing irreversible adverse health effects, or impairing an individual's ability to escape from a dangerous atmosphere. This is because it can cause eye irritation and difficulty breathing at 100 ppm. Also, unlike some other highly toxic materials, such as chlorine, H_2S is an invisible gas. Being a dense gas, it can accumulate in poorly ventilated areas. Therefore, loss of containment of H_2S is a highly hazardous event. Refineries will typically have area gas detectors for H_2S and personnel H_2S monitors that are worn when in the plant. Hydrotreaters (Section 5.10) have the role of removing sulfur and H_2S.

A refinery has many heat exchangers, distillation columns, furnaces and storage tanks, all containing flammable materials, and many containing Hydrogen Sulfide, a toxic gas. Operating problems mentioned in the sections on heat exchangers (Section 5.2), distillation columns (Section 5.3), furnaces (Section 5.6), and storage tanks (Section 5.7), can cause loss of containment, leading to fires and explosions.

Corrosion of piping and equipment, due to the impurities in crude, and use of hydrogen in several operations, is a common problem in refineries. Corrosion can be a cause of loss of containment events in any unit in a refinery. Asset Integrity and Reliability is a key PSM element for refineries.

Resources

The following sources of information about refineries have been used in the preparation of this section:

- OSHA Technical Manual – Section IV: Chapter 2 – Petroleum Refining Process, https://www.osha.gov/dts/osta/otm/otm_iv/otm_iv_2.html
- Petroleum Refining in Non-Technical Language (Ref 5.42).
- American Petroleum Institute (API) Recommended Practices (API has several recommended practice documents, some of which will be listed in the follow sections).

5.3.2 Crude Handling and Separation

Overview. Crude oil is heated in a furnace and separated into several fractions in a distillation column operating at 650 - 700 °C (1200 - 1300 °F)°. This section of the process is normally referred to as the Crude Unit. In some refineries, the heavy ends of the distillation column will go to a vacuum distillation column to further distill the crude at lower temperatures. This enables further separation at lower temperatures so that thermal cracking of the crude oil does not take place. Figure 5.39 is a process flow diagram for the atmospheric distillation column.

Example Process Safety Incidents and Hazards. The vapor cloud fire at the Chevron Richmond refinery, described in Section 2.11, occurred in the crude handling and separation unit. In that incident, a large vapor cloud fire was caused by the release of vapors from a rupture of an 8 inch line due to sulfidation corrosion. Sulfidation corrosion is due to the reaction between sulfur compounds, especially H_2S, and iron at temperatures of 230 – 430 °C (450 – 800 °F). This causes the thinning of materials such as steel, leading to failure of piping if not monitored and controlled.

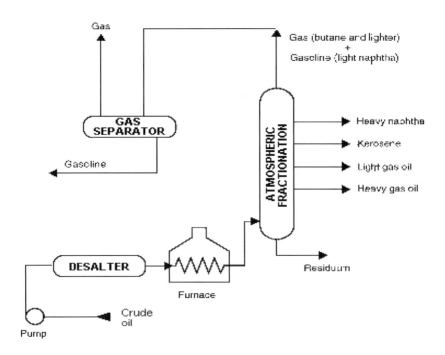

Figure 5.39. Atmospheric separation process flow diagram, courtesy OSHA.

PROCESS SAFETY IN DESIGN   183

The presence of impurities in crude such as hydrogen sulfide (H_2S) and other sulfur compounds can lead to sulfidation corrosion in all parts of the crude unit. The hazard can be reduced by the use of steel with higher chromium content. Such steels are inherently safer than carbon steel with respect to sulfidation corrosion. Ammonia can be injected into the column to control corrosion. A good Mechanical Integrity program is still required to manage the corrosion hazards. API has a Recommended Practice, API RP 939-C *Guidelines for Avoiding Sulfidation (Sulfidic) Corrosion Failures in Oil Refineries*, First Edition, 2009.

5.3.3 Light Hydrocarbon Handling and Separation

Overview. Gases from the crude separation unit are sent to a gas handling unit, sometimes called a sat gas plant (sat stands for saturated, i.e. the carbon atoms are fully saturated with hydrogen). A typical process involves compression to liquefy the gases, phase separation, absorption of the gas with lean oil, followed by a series of fractionating columns to separate ethane, propane and butanes.

Example Process Safety Incidents and Hazards. Although the Esso Longford explosion described in Section 3.5 was not in a refinery, the incident is indicative of one of the main hazards in a gas plant, auto-refrigeration. Auto-refrigeration can occur on adiabatic expansion of gasses. The resulting low temperature can bring metals like carbon steel below their ductile-brittle transition temperature resulting in metal embrittlement. This has resulted in complete rupture of vessels and pipelines with loss of containment and gas explosions. The condensation of moisture on a propane tank for a propane gas grill is an example of auto-refrigeration. Auto-refrigeration is a potential problem in chemical as well as petrochemical processes. In addition to LPGs, gases such as ammonia, chlorine and hydrogen chloride can cause auto-refrigeration.

AIChE offers a course on Auto Refrigeration and Metal Embrittlement. This course is free to undergraduate students at:
http://www.aiche.org/ccps/resources/chemeondemand/conference-presentations/auto-refrigeration-and-metal-embrittlement.

In pumps, the light hydrocarbons (LHCs) are under pressure and pump seals create a loss of containment hazard. More substantial seals, such as dual seal with barrier fluid, will typically be used in pumps in the sat gas plant.

The LHCs are under pressure. Failure of back pressure valves, which are meant to reduce pressure, can lead to vaporization in the unit and higher pressures that can cause catastrophic rupture of piping and equipment. This can lead to large flammable releases. Pressure relief valves routed to flare systems are typically installed to prevent overpressure.

External fires in the sat gas plant can lead to BLEVEs, described in Section 5.7.3.

5.3.4 Hydrotreating

Overview. The main purpose of hydrotreating is to remove impurities such as sulfur, nitrogen, oxygen and metals. Feed is mixed with hydrogen, preheated to 600 – 800 °C (1100 – 1472 °F), and charged at high pressures (up to 69 Bar (1,000 psi)) to a catalytic reactor to form H_2S, ammonia (NH_3) and metal chlorides. The petroleum portion of the feed, containing olefins and aromatics, reacts with hydrogen to form saturated compounds. The product stream is depressurized and cooled. Excess hydrogen is recycled, and the rest of the stream is sent to a column to remove butanes from the naphtha product. Figure 5.40 is a process flow diagram of a hydrotreating unit. As can be seen in Figure 5.38, there can be several hydrotreaters in a refinery.

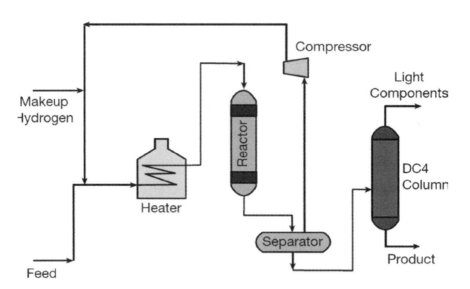

Figure 5.40. Hydrotreater process flow diagram, Ref. 5.43.

Example Process Safety Incidents and Hazards. An explosion occurred in the hydrotreating section of the Tesoro Anacortes Refinery in 2010 (Ref. 5.44). In this incident, a heat exchanger ruptured, releasing hydrogen and naptha at 500 °C (930 °F), which ignited and caused a fire that killed seven people.

A CSB video of this event is available at www.csb.gov/videos/. The rupture was due to a phenomenon called High Temperature Hydrogen Attack (HTHA).

In HTHA, hydrogen diffuses into the steel walls of equipment at high temperatures and reacts with carbon in the steel, producing methane. This reduces the carbon in the steel, causing pressure inside it. The methane causes fissures to form on the steel, weakening it. The heat exchangers at the Tesoro refinery were carbon steel, which is susceptible to HTHA. HTHA is difficult to identify in its early stages, as the fissures are very small. By the time it can be detected, the equipment already has a higher likelihood of failure. High chromium steel is more resistant to HTHA and is, therefore, a safer material of construction.

The American Petroleum Institute (API) has a recommended practice regarding HTHA; API RP 941, *Steels for Hydrogen Service at Elevated Temperatures and Pressures in Petroleum Refineries and Petrochemical Plants*, 7th Edition, 2008. API 941 provides a curve (called Nelson curve) that shows the temperatures and pressures at which HTHA can occur for various metals. The CSB investigation found that the Nelson curve was inaccurate, and API issued an alert to that effect in 2011.

Another process hazard is the potential for reverse flow of high pressure hydrogen from the hydrotreater to the upstream process if forward liquid feed flow is lost (e.g., feed pumps trip off). Check valves and/or chopper valves, which are meant to prevent reverse flow, are used to reduce the risk of this happening.

The hydrotreating reaction is exothermic and controlled by maintaining the proper feed rates and temperatures for the composition of the feed. Loss of control can lead to excess heat generation and higher than normal temperatures. The high temperatures can weaken the vessels and potentially lead to loss of containment. Operators can try to control the reaction by adjusting feed rates or preheat temperatures. Quench systems can be installed as a safeguard.

5.3.5 Catalytic Cracking

Overview. Catalytic cracking uses a catalyst to break down heavy fractions from the crude distillation unit into lighter ones such as gasoline and kerosene. The most common process is Fluid Catalytic Cracking (FCC). Figure 5.41 is a process flow diagram of an FCC. The FCC unit is one of the largest physical units in a refinery. The flow oil and catalyst mix in a riser at 425 – 480 °C (800 – 900 °F), where the reactions take place. The catalyst is separated from the product in a disengagement chamber and goes to the regenerator where it is regenerated by adding air to burn off coke that has formed on it. Catalyst exit temperatures are 650 – 815 °C (1,200 – 1,500 °F). The regenerated catalyst is then returned to the riser of the reactor.

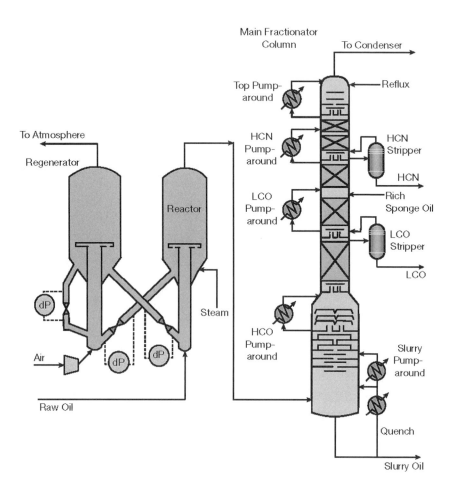

Figure 5.41. Fluid Catalytic Cracking (FCC) process flow diagram, Ref. 41.

The product flows to a column where product fractions are separated. The slurry oil is recycled back to the reactors.

Process Safety Hazards. Erosion of the piping by the catalyst can lead to loss of containment. Inspections need to check for leaks due to erosion.

Reverse flow through the reactor slide valves (see the dP cells in Figure 5.41) can result in air introduction into the reactor. This can lead to a flammable mixture, and ignition of the hot hydrocarbons, resulting in a fire and or explosion.

PROCESS SAFETY IN DESIGN   187

Removing spent catalyst is potentially hazardous due to the potential for fires from iron sulfide formation. The coked catalyst must be cooled and wetted before being dumped into containers.

5.3.6 Reforming

Overview. Reforming is the process used to convert naphthenes and paraffins to aromatics and isoparaffins, increasing the octane rating. The process also releases hydrogen, which is used in the hydrotreaters. There are two main designs, semi-regeneration and continuous catalyst regeneration (CCR). A CCR process flow diagram is shown in Figure 5.42. The reactor unit is actually a series of reactors

Process Safety Hazards. The reforming section is also subject to HTHA. Also, the hydrogen may combine with chlorine compounds to form hydrogen chloride, leading to chloride corrosion. A good inspection program is needed to check for leaks due to corrosion. Emissions of carbon monoxide and hydrogen sulfide can occur during the catalyst regeneration.

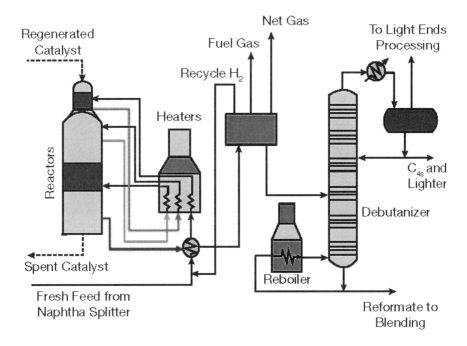

Figure 5.42. CCR Naphtha Reformer process flow diagram, Ref. 43.

5.3.7 Alkylation

Overview. Alkylation units react isobutene with propylene and butylene to create alkylate, which is a mixture of high octane materials such as isooctane. Sulfuric acid (H_2SO_4) or hydrofluoric acid (HF) is used as a catalyst. Figure 5.43 is a flow diagram of an alkylation unit.

In sulfuric acid catalyzed alkylation units a chiller reduces the petroleum feed to about 4-5 °C (40 °F), and then the feed is mixed with the acid catalyst in the reactor. Acid is then separated and recycled to the reactor in a settler. A series of fractionators separate propane, butanes and the alkylate.

Example Process Safety Incidents and Hazards. The alkylation unit uses large volumes of sulfuric acid or HF. Both are corrosive and highly hazardous. Loss of containment is a hazardous event. Loss of HF, in particular, is highly hazardous. In addition to being highly corrosive, HF is toxic. Absorption through the skin can cause cardiac arrest and inhalation causes damage to the linings of the lungs. HF can form a cloud that can travel outside of a refinery, as happened in the Texas City event described below. Some units will have automatic systems to detect a release and spray large amounts of water on an HF release to remove or scrub it from the air.

HF release, Texas City, TX, 1987. A crane lifting a heat exchanger failed causing it to drop the exchanger which severed a 4 inch loading line and 2 inch pressure relief line of an alkylation unit settling drum containing HF plus isobutane. The drum was under pressure and about 18,000 Kg of HF and 17,900 Kg of isobutylene were released. Concentrations of HF of 50 ppm were noted about three quarters of a mile from the source of the release, based on damage to vegetation. That level of HF is considered to be the threshold level above which life threatening effects can be observed. The release was mitigated by transferring as much HF as possible from the settler to railcars and by spraying water on the release. It took 44 hours to stop the release. (Ref. 5.45)

HF release, Corpus Christi, TX, 2009. A control valve failed closed, blocking flow in process piping. The sudden flow blockage caused violent shaking in the piping, which broke two threaded connections. There was a release of flammable hydrocarbons which ignited. The fire caused several other failures, including the release of about 42,000 pounds (19050 Kg) of HF. There was a water mitigation system, which did activate and absorbed most of the HF. One worker was critically burned. Citgo reported that 30 pounds of HF were not captured by the water mitigation system. Studies have shown that the best these systems can do is 95% removal efficiency. The CSB recommendations state that, at 90% efficiency, the atmospheric release would have been about 4,000 pounds (1814

PROCESS SAFETY IN DESIGN 189

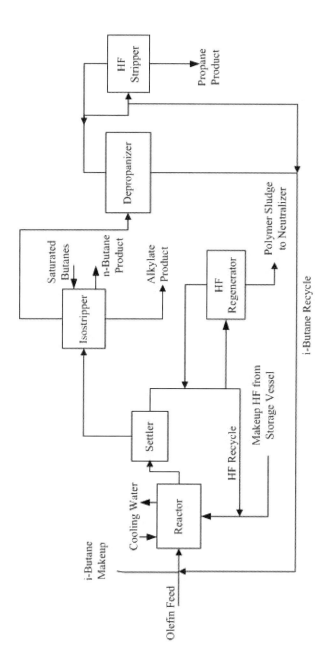

Figure 5.46. HF Alkylation process flow diagram. Ref. 5.46.

Kg). The water supply for the system was nearly used up and salt water from the adjacent ship channel was used for firefighting. The CSB found that Citgo had never conducted a safety audit of the unit (Ref. 5.47). The API practice regarding alkylation, listed below, recommends a safety audit every three years

The alkylation reaction is exothermic. Loss of control has the same consequences as in the hydrotreator.

The API has a recommended practice regarding HF Alkylation; API RP 751, *Safe Operation of Hydrofluoric Acid Alkylation Units*, Third Edition, June 2007.

5.3.8 Coking

Overview. The Coker takes the heavy feedstock and thermally cracks them to produce lighter products. The residue is a solid called Coke. There are two main coking processes, Delayed Coking and Continuous Coking. Figure 5.47 is the flow diagram for a delayed coking unit.

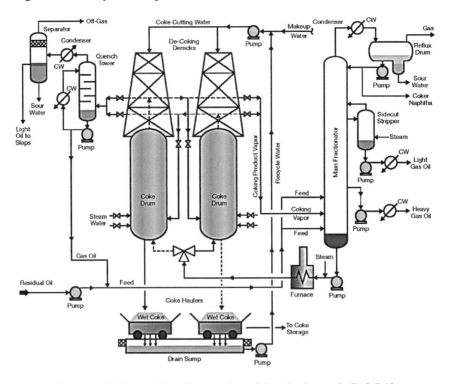

Figure 5.44. Process flow diagram for a delayed coker unit, Ref. 5.43.

The bottoms from the crude distillation unit are heated to about 450 - 500 °C and charged to the bottom of a coke drum. The material sits in the drum for about 24 hours at 3 – 8 bar (40 – 115 psig) and thermal cracking to form lighter products continues. The lighter product is drawn off to a fractionator and recovered material sent to other parts of the refinery for processing. When a predetermined level is reached, flow is directed into a second coke drum. Some refineries have several coke drums. Coke drums can be up to 37 m (120 feet) tall and 9 m (29 feet) in diameter. The first drum has to be cooled, the tops and bottoms removed, and then the drum washed with high pressure fluids to remove the coke.

Example Process Safety Incidents and Hazards.

Equilon Anacortes Refinery Coking Plant Accident, 1998. A storm caused a power interruption during the first hour of filling a drum. The charge line itself became clogged with coke. Operators tried to clear the pluggage by steaming out the line, and believed they had succeeded. Based on temperature readings the staff concluded cooling of the drum was done. The top head was removed, and then the bottom head. As the bottom head was removed, hot heavy oil broke through a crust and ignited because it was above its auto-ignition temperature. There was an explosion and fire. Six people were killed. The staff was misled by temperature sensors located on the outside of the drum instead of in it. Equilon subsequently installed a remote controlled cleaning system (Ref. 5.48).

Delayed Cokers have been a source of many serious accidents and were the subject of an OSHA Safety Hazard Information Bulletin (SHIB) on *Hazards of Delayed Coker Unit (DCU) Operations* (Ref. 5.49).

If the switching valves are not properly aligned, or are leaking through, hot material can be sent to the drum being cleaned, leading to loss of containment and potential fires and explosion. Opening the wrong valve has "led to serious incidents" per the OSHA SHIB. Providing interlocks to control valve opening can prevent this from occurring.

In some delayed coking units the cleaning is done manually. When the coke drum heads are removed, workers can be exposed to geysers of steam, hot water, coke particles, hot tar balls through the top drum and avalanches of coke from the bottom head. It is difficult to predict when material can be ejected from the drum heads. Operator training to be prepared for the hazards of opening the drums is needed. Shrouds around the drum head or an automated removal system can mitigate the hazard. Some units use remote controlled cleaning units to avoid exposing operators to potential hazards.

With some feeds, foaming can occur causing high drum pressure and level and plugging of the drum outlets and relief valves. This can result in overpressurization and loss of containment. Antifoams can be added to these feeds to prevent this.

5.4 Transient Operating States

5.4.1 Overview

Transient operations include normal startups, startups after an emergency shutdown, normal and emergency shutdowns, extended holds for maintenance and recovery from process upsets or emergency operations. In some shutdowns, part of a process may continue to run in a holding mode until ready for restart. Startups and shutdowns are analogous to takeoffs and landings in the aerospace industry. These operating phases usually involve more operator intervention with a process than the continuous or normal operation mode, and are therefore more subject to human error. Additionally, many engineered safety systems, such as automated interlocks or controls may be not be meaningful during the period during startup and shutdown. Some of these automatic systems can be programmed to start when process conditions reach specified settings.

5.4.2 Example Process Safety Incidents

Catch Tank Explosion. An explosion occurred in a BP Amoco plant in 2001 after a startup attempt for an extruder had to be aborted. During normal startup polymer was directed from a reactor to a catch tank, Figure 5.48, until the extruder could be started up and was running normally. On this occasion there was difficulty starting the extruder, so the startup was aborted. About 12 hours later operators on the night shift were told to remove the cover from the catch tank for cleaning. While attempting to remove the cover it blew off the catch tank when about half the bolts were removed. Three maintenance workers were killed (Ref. 5.50).

During the startup attempt, more than twice the normal amount of polymer was sent to the catch tank. This included polymer and flush solvent. Because the catch tank contained more polymer than usual, the vapor that entered with it could not escape. The vent system was likely blocked by polymer allowing pressure and temperature to build up in the catch tank. The polymer also blocked the pressure indicator, so no one knew of the high pressure in the tank. This pressure was released violently as the maintenance workers began to open the tank.

PROCESS SAFETY IN DESIGN 193

Figure 5.45. Polymer catch tank, Ref. 5.50.

Industry experience shows that the frequency of incidents is higher during process transitions such as startups or restarts from temporary idle conditions. Several of the incidents described in previous chapters occurred during a transient operating mode. They are shown in the list below. The fact that seven of the sixteen incidents described in Chapter 3 are in this list is an indication of the risk of transient operations:

- NASA Space Shuttle Challenger explosion (Section 2.2)
- Motiva Refinery explosion (Section 2.10)
- Hydrocracker explosion (Section 2.16)
- BP Refinery explosion, Texas City (Section 3.1)
- ARCO Channelview explosion (Section 3.2)
- NASA Space Shuttle Columbia Disaster (Section 3.3)
- Esso Longford gas plant explosion (Section 3.5)
- Port Neal AN explosion (Section 3.6)
- Texaco, Milford Haven, UK, explosion (Section 3.9)
- Macondo well blowout (Section 3.16)
- Distillation column incident (Section 5.3.2)
- Carbon bed incident (Section 5.3.2)
- Seveso (Section 5.5.2)
- Fired equipment (Section 5.6.2, examples 1 and 3)

- Equilon Anacortes Refinery Coking Plant Accident (Section 5.14.2)

5.4.3 Design Considerations

Since engineered safety systems are often offline during transient operations, the role of operators and technical personnel, and their knowledge of the process operation, are critical. Written procedures are required for these operating modes, in fact, these are required for processes covered by the OSHA PSM and EPA RMP standard. Emergency or abnormal procedures must include what actions operators should take when process conditions go beyond their defined limits.

Risk associated with transient operating states should be identified in the HIRA. HIRAs must include startup and shutdown, loss of utilities, and should define the responses to the identified process upsets. This information can be turned into emergency procedures. The PHA can be used to document the risk in transient operations, and write operating procedures, providing adequate training and refresher training on the risk when startups and shutdowns occur. Operational readiness reviews should be done before startups as described in Section 2.15. Management of change reviews must be held when unusual/extended holds occur, as in the Arco Channelview and Seveso incidents.

5.5 References

5.1. Inherently Safer Chemical Processes; A Life Cycle Approach, 2nd Ed. American Institute of Chemical Engineer, New York, NY, 2009.
5.2. Kletz, T., Process Plants: A Handbook for Inherently Safer Design, Taylor and Francis, London, 1998.5.1 Guidelines for Engineering Design for Process Safety (Second Edition), Center for Chemical Process Safety, New York, 2012.
5.3 ASME Section VIII-DIV 1. *ASME Boiler and Pressure Vessel Code,* Section VIII, Division 1: Rules for Construction of Pressure Vessels, American Society of Mechanical Engineers, New York, NY, 2010.
5.4 CCPS, Process Safety Beacon, The Seal that Didn't Perform, July 2002, (http://sache.org/beacon/files/2002/07/en/read/2002-07%20Beacon-s.pdf)
5.5 CCPS, Process Safety Beacon, It's a Bird, It's a Plane, It's ..A Pump, October 2002 (http://sache.org/beacon/files/2002/10/en/read/2002-10%20Beacon-s.pdf)
5.6 Kelley, J. Howard, Understand the Fundamentals of Centrifugal Pumps, Chemical Engineering Progress, p 22-28, Oct 2010.
5.7 Berg, J. The Case for Double Mechanical Seals, Chemical Engineering Progress, p. 42-45, June 2009.
5.8 CCPS Process Safety Beacon, Understand the Reactivity of Your Heat Transfer Fluid, February 2011 (http://sache.org/beacon/files/2011/02/en/read/2011-02-Beacon-s.pdf)
5.9 Mukherjee, R., Effectively Design Shell-and-Tube Heat Exchangers, Chemical Engineering Progress, Feb. 1998.

5.10 Haslego and Polley, Designing Plate-and Frame Heat Exchangers, Chemical Engineering Progress, p. 30-37, Sept. 2002.
5.11 Chu, Chu, Improved Heat Transfer Predictions for Air-Cooled Heat Exchangers, Chemical Engineering Progress, p. 46-48, Nov. 2005.
5.12 API STD 660. Shell-and-Tube Heat Exchangers, Eighth Edition, American Petroleum Institute, Washington, DC., 2007.
5.13 Bouck, Doug, Distillation Revamp Pitfalls to Avoid, Chemical Engineering Progress, p. 32-38, Feb. 2014.
5.14 Ender, Christophe and Laird, Dana, Minimize the Risk of Fire During Column Maintenance, Chemical Engineering Progress, p. 54-56, September 2003.
5.15 Mannan, Sam, Best Practices in Prevention and Suppression of Metal Packing Fires, Mary Kay O'Connor Process Safety Center, August 2003.
5.16 OSHA Safety Hazard Information Bulletin on Fire Hazard from Carbon Adsorption Deodorizing Systems, August 17, 1992.
(https://www.osha.gov/dts/hib/hib_data/hib19970730.html)
5.17 Naujokas, A.A., Spontaneous Combustion of Carbon Beds, *Plant/Operations Progress*, p. 120-126, April 1995.
5.18 Jarvis, H.C. Butadiene Explosion at Texas City-2, *Plant Safety & Loss Prevention*, Vol. 5. 1971.
5.19 Butadiene Explosion at Texas City-1, *Plant Safety & Loss Prevention*, Vol. 5.
5.20 Keister, R.G., et al. Butadiene Explosion at Texas City-3, *Plant Safety & Loss Prevention*, Vol. 5. 1971.

5.21 Sherman, R.E., Carbon-Initiated Effluent Tank Overpressure Incident, *Process Safety Progress*, Vol. 15, No. 3, p. 148-149, Fall 1996.
5.22 Guidelines for Safe Handling of Powders and Bulks Solids, Center for Chemical Process Safety, New York, 2005.
5.23 Drogaris, G. Major Accident Reporting System: Lessons Learned from Accidents Notified, Elsevier Science Publishers, B.V, Amsterdam, 1993.
5.24 Kletz, T, What Went Wrong, Case Histories of Process Plant Disasters, 4th Ed., Elsevier, Houston, TX, 1993.
5.25 Garland, R. Wayne, Root Cause Analysis of Dust Collector Deflagration Incident, *Process Safety Progress*, Vol. 29, No. 4, December 2010.
5.26 Patnaik, T., Solid-Liquid Separation: A Guide to Centrifuge Collection, Chemical Engineering Progress, p. 45-50, July 2012.
5.27 Lees Loss Prevention in the Process Industries, Vol. 3, Elsevier, 2012. ISBN978-0-12-397212-5.
5.28 U.S. Chemical Safety and Hazard Investigation Board, Investigation Report, Report No. 2008-3-I-FL, T2 Laboratories, Inc. Runaway Reaction. Jacksonville, FL. September 2009. (http://www.csb.gov/investigations).
5.29 CCPS Process Safety Beacon, Interlocked for a Reason, June 2003. (http://sache.org/beacon/files/2003/06/en/read/2003-06%20Beacon-s.pdf)
5.30 CCPS Process Safety Beacon, Avoid Improper Fuel to Air Mixtures, Jan. 2004. (http://sache.org/beacon/files/2004/01/en/read/2004-01%20Beacon-s.pdf)

5.31 Ramzan, Naveeed, et al, Root Cause Analysis of Primary Reformer Catastrophic Failure: A Case Study, *Process Safety Progress*, Vol. 30, No. 1, p 62-65, March 2011.

5.32 The Buncefield Incident, The final report of the Major Incident Investigation Board, Volume 1, 11 December 2008. (http://www.hse.gov.uk/comah/buncefield/miib-final-volume1.pdf)

5.33 CCPS Process Safety Beacon, The Great Boston Molasses Flood of 1919, May 2007. (http://sache.org/beacon/files/2007/05/en/read/2007-05-Beacon-s.pdf)

5.34 CCPS Process Safety Beacon, What if You Load the Wrong Material Into a Tank?, April 2012. (http://sache.org/beacon/files/2012/04/en/read/2012-04-Beacon-s.pdf)

5.35 CCPS Process Safety Beacon, Vacuum is a Powerful Force!, Feb. 2002, (http://sache.org/beacon/files/2002/02/en/read/2002-02-Beacon-s.pdf)

5.36 US EPA, Operating And Maintaining Underground Storage Tank Systems, EPA 510-B-05-002, September 2005. (http://www.epa.gov/oust/pubs/ommanual.htm)

5.37 NFPA 30, Flammable and Combustible liquid Storage Code, National Fire Prevention Association, Quincy, MA, 2015.

5.38 Britton, L.G., Avoiding static ignition hazards in chemical operations, *AIChE-CCPS Concept Book*, New York, (1999).

5.39 NFPA 77, Recommended Practice on Static Electricity, National Fire Prevention Association, Quincy, MA, 2014.

5.40 NFPA 780, Standard for the Installation of Lightning Protection Systems, National Fire Prevention Association, Quincy, MA, 2014.

5.41 Urban, P.G., Bretherick's Handbook of Reactive Chemical Hazards (7th Edition), Academic Press, New York, NYISBN:978-0-12-372563-9, 2006.

5.42 Leffler, Willaim, W., Petroleum Refining in Nontechnical Language, 4th Edition, PennWell, Tulsa, OK., 2008.

5.43 Olsen, Tim, An Oil Refinery Walk-Through, Chemical Engineering Progress, Vol. 110, No. 5, p. 34-40, May 2014.

5.44 U.S. Chemical Safety and Hazard Investigation Board, Investigation Report, Catastrophic Rupture of Heat Exchanger, Report No. 2010-08-I-WA, May 2014.

5.45 Woodward, John L. and Hillary Z., Analysis of Hydrogen Fluoride Release at Texas City, *Process Safety Progress*, Vol. 17, No. 3, p. 213-218, Fall 1998.

5.46 Kaiser, Geoffrey D., Accident Prevention and the Clean Air Act Amendments of 1990 with Particular Reference to Anhydrous Hydrogen Fluoride, PSP, Vol. 12, No. 3, p. 176-180, July 1993.

5.47 U.S. Chemical Safety and Hazard Investigation, Urgent Recommendations, 12/09/2009. (http://www.csb.gov/assets/1/19/Urgent_Recommendations_to_Citgo_-_Final_Board_Vote_Copy1.pdf)

5.48 U.S. Chemical Safety and Hazard Investigation, Safety Bulletin, Management of Change, No. 2001-04-SB, Aug. 2001. (http://www.csb.gov/assets/1/19/moc0828011.pdf)

5.49 Hazards of Delayed Coker Unit (DCU) Operations, OSHA SHIB 03-08-29(C), 2003, (https://www.osha.gov/dts/shib/shib082903c.html).

5.50 U.S. Chemical Safety and Hazard Investigation Board, Investigation Report, Report No. 2001-03-I-GA, Thermal Decomposition Incident., BP Amoco Polymers, Augusta, GA, June 2002. (http://www.csb.gov/investigations).

6

Course Material

6.1 Introduction

The purpose of this chapter is to show how process safety can fit into chemical engineering courses. This chapter is a guide for instructors, as well as chemical engineering students. The process safety modules from the Safety and Chemical Engineering Education (SACHE) group are described, with suggestions as to how they can be integrated into existing courses. Access to SACHE modules is available through the SACHE website, www.sache.org. To access materials, you will need a sign in name and password. Generally, every university in the U.S. with a chemical engineering department has permission to this website. Further, many departments have a departmental member who is the contact for the SACHE website will. The full list of SACHE courses is provided in Appendix D.

The SACHE courses described are grouped by topic: Inherently Safer Design, Process Safety Management, Hazards, Hazard Identification and Risk Assessment, Protection Systems, Case Histories and Other.

6.2 Inherently Safer Design

The **Inherently Safer Design** (ISD) and **Inherently Safer Design Conflicts and Decisions** courses are excellent resources for Chemical Engineering Plant Design and Chemical Engineering Kinetics/Reactor Design courses. It also has applications to Chemistry. These SACHE courses describe what ISD is, and why we want to practice it. They also describe the hierarchy of process safety strategies listed in Section 5.1 and the four strategies for designing inherently safer processes, providing examples for each.

In a similar vein, SACHE's **Green Engineering Tutorial** discusses a methodology for more environmentally friendly designs. A software tool is available from the tutorial's author to analyze processes.

An associated SACHE resource is *An Inherently Safer Process Checklist*.

6.3 Process Safety Management and Conservation of Life

These two SACHE products are appropriate for Plant Design and Material and Energy Balances courses.

Process Safety Management Overview describes the original 12 elements of process safety put forward by the CCPS. **Conservation of Life: Application of Process Safety Management** provides an overview of the application of process safety. Conservation of Life (COL) is based on the concept that COL is a fundamental principle of chemical engineering design and practice, equivalent in importance to conservation of energy and mass. COL principles that are described include:

- **Assess material/process hazards** – find or develop data on flammability, toxicity, reactivity, etc.
- **Evaluate hazardous events** – consequence analysis is used to evaluate the undesirable effects of potentially hazardous events
- **Manage process risks** – apply inherently safer approaches, design multiple layers of protection, evaluate risk versus tolerable risk criteria
- **Consider real-world operations** – implement PSM systems, learn from experience
- **Ensure product sustainability** – implement product safety/stewardship practices

The COL course contains many examples of these principles applied to real cases.

6.4 Process Safety Overview and Safety in the Chemical Process Industries

These courses provide overviews of the field of process safety.

Process Safety Overview. This SACHE product is based on the book *Chemical Process Safety, Fundamentals with Applications* (Ref. 6.1) contains 31 presentations that cover several process safety topics. The hazards covered in this material include; toxicity and industrial hygiene, fires and explosions. There are also materials covering Hazard Identification and Risk Analysis (HIRA), fire and explosion protection systems, emergency relief systems and incident investigation. This material can be used as a supplement to Plant Design, Kinetics/Reactor Design, Thermodynamics, Heat Transfer, Momentum Transfer and Fluid Flow courses.

Safety in the Chemical Process Industries. This product is a series of 13 videos covering topics including; lab safety, personal protective equipment, process area safety features, emergency relief systems, dust and vapor explosions, and safety reviews. The various videos can supplement Unit Operations Lab, Kinetics/Reactor Design and Plant Design courses.

6.5 Process Hazards

SACHE has several courses that cover process hazards. This section is divided into three parts, Chemical Reactivity Hazards, Fires and Explosions and Other Hazards. The courses are:

- Chemical Reactivity Hazards
- Safe Handling Practices: Methacrylic Acid
- Seminar on Fires
- Fire Protection Concepts
- Explosions
- Dust Explosion Prevention and Control and Explosions
- Introduction to Biosafety
- Fundamentals of Chemical Transportation with Case Histories
- Metal Structured Packing Fires
- Properties of Materials
- Static Electricity as an Ignition Source
- Static Electricity I -- Everything You Wanted to Know about Static Electricity

6.5.1 Chemical Reactivity Hazards

This Chemical Reactivity Hazards module can be used in conjunction with Kinetics/Reactor Design, Thermodynamics, Heat Transfer and Plant Design courses.

"Safely conducting chemical reactions is a core competency of the chemical manufacturing industry." (Ref. 6.2). The CSB report this quote came from is a study of 167 chemical reactivity incidents, highlighting the importance of reactivity hazards. These are not well addressed in university training. Documentation of the process chemistry, process material's reactivity, thermal and chemical stability and the hazards of accidental mixing of materials are among the items specifically called for by the OSHA PSM regulation. Changes to the process chemistry must be studied in a Management of Change review.

This web-based instructional module contains about 100 web pages with extensive links, graphics, videos, and supplemental slides. It can be used either for classroom presentation or as a self-paced tutorial. The module shows how uncontrolled chemical reactions in industry can lead to serious harm, and introduces key concepts for avoiding unintended reactions and controlling intended reactions.

The five main sections in the module cover:

1. three major incidents that show the potential consequences of uncontrolled reactions;
2. how chemical reactions get out of control, including consideration of reaction path, heat generation and removal, and people/property/environmental response
3. data and lab testing resources used to identify reactivity hazards,
4. four approaches to making a facility inherently safer with respect to chemical reactivity hazards;
5. strategies for designing facilities both to prevent and to mitigate uncontrolled chemical reactions.

The module concludes with a ten-question informative quiz. An extensive Glossary and Bibliography are directly accessible from any page.

This module is based the book, *Essential Practices for Managing Chemical Reactivity Hazards,* (Ref. 6.3). The book describes a methodology for screening chemicals for potential reactive hazards using publically available information. A flowchart of the methodology is provided in Appendix E.

Another resource for reactive chemical issues is the software program, *Chemical Reactivity Worksheet* (CRW). The CRW can predict the hazards of mixing for several thousand chemicals. The CRW can be downloaded for free from the National Oceanic and Atmospheric Administration's website (http://response.restoration.noaa.gov/reactivityworksheet).

Safe Handling Practices: Methacrylic Acid. This SACHE module covers a specific chemical, methacrylic acid, but is applicable to polymerizable materials in general. There is a presentation and a video of the consequences of a runaway reaction in a railcar. This can be used to supplement a Thermodynamics or Heat Transfer course.

6.5.2 Fires and Explosions

The following SACHE modules can be used in courses such as Plant Design, Thermodynamics, or a Process Safety course.

Seminar on Fires. This presentation covers the fundamentals of fires and explosions including such topics as:

- technical definition of fires and explosions,
- physical characteristics of various fires,
- necessary conditions for fires and explosions, and
- elementary properties, such as flammability limits (LFL and UFL), minimum ignition energy (MIE), flame speeds, burning rates, etc.

Fire Protection Concepts. This course consists of two sections. Section 1 covers the fundamentals of fire. Section 2 describes fire protection methods such as separation of process areas from storage, diking and impoundment, fireproofing of structures and fire extinguishment.

Explosions. This is a video that provides pictures showing the consequences of explosions and covers the fundamentals of explosions and some practices necessary for preventing explosions.

Properties of Materials. This presentation covers flammability, explosive, reactivity and toxicological properties. It shows what properties can be found on safety data sheets.

Dust Explosion Prevention and Control. This course is divided into three sections. The first one describes the conditions and consequences of dust explosions. The second part consists of videos showing dust explosions and design methods for preventing them. The third part describes the role of static electricity in dust explosions.

6.5.3 Other Hazards

Introduction to Biosafety. This module is intended to provide a brief overview of the area of biosafety. It provides an introduction to types of biohazards and discusses sources of biohazards, classifications of biohazards by risk group, and methods of reducing risk from biohazards. The module is oriented towards dealing with biohazards in a laboratory or clinical setting. Examples of Biosafety Manuals are included. This module can supplement a lab or Plant Design course.

Fundamentals of Chemical Transportation with Case Histories. This overview of transportation of chemical materials addresses transportation regulations, and the hazards of various means of transportation. Several case histories are included. This module can be used in a Plant Design or Process Safety course.

Metal Structured Packing Fires. Metal structured packing fires are a hazard unique to packed columns (see 5.3.1). This can be a supplement to a Mass Transfer course.

Static Electricity. Static electricity is the ignition source for over 10% of fires and explosions. The SACHE courses **Static Electricity as an Ignition Source** and **Static Electricity I -- Everything You Wanted to Know about Static Electricity** and the process safety overview modules mentioned above cover how static electricity is generated and discharged and how to control it. These presentations can supplement a Plant Design and Unit Operations Lab course.

6.6 Hazard Identification and Risk Analysis

SACHE Courses covering HIRA are:

- Process Hazard Analysis: An Introduction

- Process Hazard Analysis: Process and Examples
- Dow Fire and Explosion Index (F&EI) and Chemical Exposure Index (CEI) Software
- Layer of Protection Analysis
- Risk Assessment
- Safety Guidance in Design Projects
- Project Risk Analysis (PRA): Unit Operations Lab Applications
- Consequence Modeling Source Models I: Liquids & Gases
- Understanding Atmospheric Dispersion of Accidental Releases

The first set of courses on hazard analysis and risk assessment are meant to be supplements to a Plant Design course. The **PRA** course has been developed to apply to a Unit Operations Lab. The last two courses can be supplements to courses on Material and Energy Balances, Fluid Flow/Momentum Transfer and Thermodynamics.

Process Hazard Analysis: Introduction / Process and Examples. The introduction covers the definition of PHA and some basic hazards. Included is information that can be used as an exercise. The second part goes into more detail on types of hazard analysis techniques, such as Hazard and Operability (HAZOP) and Failure Mode and Effects Analysis (FMEA) with several examples.

Dow F&EI and CEI Software. The Dow F&EI is a semi-quantitative measurement that provides an evaluation of the fire, explosion, and reactivity potential of process equipment and its contents. The CEI provides a method of rating the relative acute health hazard potential to people in neighboring plants or communities from possible chemical release incidents.

Layer of Protection Analysis (LOPA). LOPA is a semi-quantitative tool for analyzing and assessing risk. This technique includes simplified methods to characterize the consequences and estimate the frequencies of undesired consequences. LOPA generally employs more rigor than what is encountered with qualitative risk assessments, while still not becoming overly onerous when compared to detailed Quantitative Risk Assessments (QRA). A distillation column example is provided.

Risk Assessment. This is a web browser-based, self-study course designed to provide a working knowledge of risk assessment, management and reduction as applied to chemical plants and petroleum refineries. It includes descriptions of methods with examples and exercises, and it requires about three hours to complete.

Safety Guidance in Design Projects. This module covers how to implement hazard evaluation, risk and risk reduction strategies in a design project. The module outlines the design project steps and lists SACHE courses that are tied to the steps.

Project Risk Analysis (PRA): Unit Operations Lab Applications. This course helps a lab instructor apply PRA in a unit operations laboratory setting. Based on an industrial risk analysis approach, students document that they understand the potential hazardous events related to their project before experimental work begins based on an area tour; blank PRA check lists are provided.

Consequence Modeling Source Models. Source models are used to estimate the rates and quantities of material released when an incident occurs. This course is an introduction of source models used to model releases in the risk assessment process. Source modeling involves Thermodynamics, and Momentum Balance/Fluid Flow and Material and Energy Balances Example problems are provided.

Understanding Atmospheric Dispersion of Accidental Releases. This is a short (~ 50 page) concept book that provides an introduction to dispersion modeling. Dispersion modeling is used to calculate the concentration of gases, vapors and aerosols that travel downwind after a release (source modeling) occurs.

6.7 Emergency Relief Systems

The design of Emergency Relief Systems (ERS) involves Momentum Balances/Fluid Flow and Thermodynamics. Design of an ERS is a part of Kinetics/Reactor Design and Plant Design. The following products are available from SACHE:

- Venting of Low Strength Enclosures
- Compressible and Two-Phase Flow with Applications Including Pressure Relief System Sizing
- Design for Overpressure and Underpressure Protection
- Emergency Relief System Design for Single and Two-Phase Flow
- Runaway Reactions -- Experimental Characterization and Vent Sizing
- Safety Valves: Practical Design Practices for Relief Valve Sizing
- Simplified Relief System Design Package
- University Access to SuperChems and ioXpress

Venting of Low Strength Enclosures. This course describes how to vent low strength enclosures, i.e. buildings, to protect them from internal explosions and shows the consequences of unvented explosions.

Compressible and Two-Phase Flow with Applications Including Pressure Relief System Sizing. This product provides three Microsoft Excel programs for fluid flow calculations. It covers mass, momentum, and energy balances for fluid flow in pipes and orifices. These materials can be used in several courses; e.g., fluid mechanics, heat transfer, and senior design courses.

Two of these courses; **Simplified Relief System Design Package** and **University Access to SuperChems and io Express**, provide connections to free software for sizing emergency vents.

The rest of these courses cover emergency relief scenarios, relief devices and how to size ERS. Several come with Microsoft Excel software examples.

6.8 Case Histories

The following is a list of fourteen case histories available through SACHE. A few cover incidents already described in Chapters 2 and 3. A brief description of the incident is provided along with appropriate courses which they could supplement:

- Case History: A Batch Polystyrene Reactor Runaway & Mini-Case Histories: Monsanto
- Rupture of a Nitroaniline Reactor
- T2 Runaway Reaction and Explosion
- Mini-Case Histories: Morton
- Seminar on Tank Failures
- The Bhopal Disaster: A Case History & Mini-Case Histories: Bhopal
- Hydroxylamine Explosion Case Study
- Piper Alpha Lessons Learned
- Seveso Accidental Release Case History
- Mini-Case Histories: Flixborough
- Mini-Case Histories: Hickson
- Mini-Case Histories: Phillips
- Mini-Case Histories: Sonat
- Mini-Case Histories: Tosco

The **Mini-Case Histories** module contains eight case histories. In two cases, Bhopal and the Polystyrene Runaway, there is also separate case history module on the incident. These newer modules are recommended over the mini-case history modules.

6.8.1 Runaway Reactions

The first four cases are runaway reactions. Each of these can be used to supplement a Kinetics/Reactor Design, Thermodynamics, Heat Transfer or Process Control course. All of these were batch reactors in which all of the reactants were present at the start of the batch, and, as such, are good examples of the potential to apply ISD principles by using a semi-batch or continuous process.

Batch Polystyrene Reactor Runaway. In this process all of the styrene was charged to the reactor and the batch heated to 95 °C and held for 2 – 8 hours. A temperature controller was used to run the reactor. The batch in question was overheated due to a temperature controller error. Styrene vapors escaped into the process building through a failed sight glass and ignited, causing an explosion that destroyed the process building, overturned a railcar and killed 11 employees.

Rupture of a Nitroaniline Reactor. A mischarge in a reaction to form nitroaniline ($O_2NC_6H_4NH_2$) caused the reaction to proceed at too high of a rate. The heat generated was not removed rapidly enough causing the reactor contents to increase in temperature further accelerating the rate of reaction. Materials containing a nitro group ($-NO_2$) can decompose violently when overheated. Within minutes, the temperature became high enough that decomposition of the nitroaniline occurred which ruptured the reactor, injured four people, one severely, destroyed the process building, and broke windows in a building three miles away.

T2 Laboratories runaway reaction. See Section 5.5.2. The reactor's emergency relief system was not capable of relieving the pressure from the runaway reaction.

Morton Reactor Runaway. A dye was made by adding ortho-nitrochlorobenzene and 2-ethylhexylamine to a reactor and manually heating the contents until the reaction was initiated, then cooling was applied. On the batch in question, the reactants were heated too rapidly, leading to a runaway reaction. The investigation revealed that the cooling system was inadequate for the reaction. The ERS was too small to safely vent the reaction, so the pressure increase blew the manway off the vessel, releasing the contents into the building and leading to a fire and explosion.

6.8.2 Other Case Histories

Seminar on Tank Failures. This module can be a supplement to course in Plant Design and Heat Transfer. The presentation covers three storage tank case histories; BLEVEs (see Section 3.14 Mexico City LPG explosions and Section 5.7.3 Storage - Design Considerations), failure of a Liquefied Natural Gas (LNG) tank and failure of a diesel storage tank.

The Bhopal Disaster: This event is described in Section 3.15. The Bhopal incident can supplement a Plant Design course. It can also be used to illustrate the benefits of ISD. The plant design could have *minimized* the amount of methyl isocyanate (MIC) stored or the chemistry could have been modified to *substitute* a less hazardous material than MIC. The presentation shows an alternative chemistry that makes a different, safer, intermediate product than MIC.

Hydroxylamine Explosion Case Study. This incident is described in Section 3.4. The module provides slides to present the case study in a course on Plant Design.

Piper Alpha Lessons Learned. This incident is described in Section 3.7. The case study can supplement a course on Plant Design.

Seveso Accidental Release Case History. This incident is briefly described in Section 2.1. Leading up to the runaway reaction at Seveso, operators had cooled the batch to 158 °C, well below the onset of decomposition of 230 °C. The reactor walls above the batch, however, were closer to 300 °C from contact with the superheated steam. Heat transfer calculations could have shown there was enough energy available in the heated metal to raise the surface of the liquid to above the decomposition onset. Consequence modeling could have shown the extent of a release and led to a better ERS design. The case study can supplement courses on Material and Energy Balances, Heat Transfer, Thermodynamics, Kinetics/Reactor Design, Plant Design.

Mini-Case Histories: Flixborough. This incident was described in Section 3.11. It can be used as a supplement to a Plant Design course as a lesson to bring in people with the correct expertise when designing piping systems, especially those that run at elevated temperatures and pressures.

Mini-Case Histories: Hickson. An explosion occurred in a vacuum distillation unit which recovered isopropyl alcohol (IPA) from a mother liquor. The incident occurred when a power outage caused loss of cooling and agitation, leading to runaway decomposition reaction involving nitro compounds in the mother liquor, which increased the pressure of the still pot, causing it to rupture. The explosion caused one major injury, damaged equipment, and destroyed the process building. This module can supplement a course on Kinetics/Reactor (understanding potential reactions in the mother liquor and ERS sizing), and Heat Transfer and Thermodynamics (to calculate how long it would take for the temperature to increase above its decomposition onset temperature).

Mini-Case Histories: Phillips. In 1989 an explosion and fire occurred at the Phillips 66 Company Houston Chemical Complex in Pasadena, Texas. 23 workers were killed, and more than 130 were injured. Property damage was over 500 million dollars. Workers were manually cleaning a settling leg on a High Density Polyethylene reactor. A single block valve used to isolate the settling leg from the reactor was opened due to an error in air supply connection to the valve and the reactor contents were dumped out, ignited and exploded. This event can supplement a course in Reactor or Plant Design, a better designed system, such as double block and bleed valve, could have prevented this event.

Mini-Case Histories: Sonat. At a petroleum separation facility a vessel was overpressurized and burst, leading to a fire and four fatalities. The overpressurization occurred when a manual valve alignment allowed a high pressure to build up in a vessel not rated for it. The design was never reviewed. This module can supplement a Plant Design or Process Safety course.

Mini-Case Histories: Tosco. A fire occurred in the crude unit at Tosco Corporation's Avon oil refinery in Martinez, California. Workers were attempting to replace piping attached to a 150-foot tall fractionator tower while the process unit was in operation. During the removal of the piping, a leaking block valve led to a release of naphtha onto the hot fractionator and it ignited. The flames engulfed five workers located at different heights on the tower. Four workers were killed, and one sustained serious injuries. No safety reviews were conducted for this work. This module can supplement a Plant Design or Process Safety course.

COURSE MATERIAL

6.9 Other Modules

Improving Communication Skills. This module is designed to supplement junior and senior chemical engineering courses in which written or oral reports about experiments or other assignments are part of the course. This can specifically supplement a Unit Operations Lab course.

Student Problems. "Safety, Health, and Loss Prevention in Chemical Processes - Problems for Undergraduate Engineering Curricula, Volume 1" was originally published by CCPS in 1990, and "Safety, Health, and Loss Prevention in Chemical Processes - Volume 2" was originally published by CCPS in 2002 and distributed to SACHE University Members. The problems were designed for use in existing engineering courses, such as Stoichiometry, Thermodynamics, Fluid Mechanics, Kinetics, Heat Transfer, Process Dynamics and Control, Computer Solutions, and Mass Transfer. The authors believed that including these problems in a required undergraduate course helps engineering students develop a safety culture and mindset that will benefit them throughout their careers. Both books are out of print. These modules provide links to the problems and solutions.

Jeopardy Contest. This SACHE product contains important elementary concepts in chemical process safety. The understanding of these concepts is assessed and reinforced with two class Jeopardy Games. For the game, it is recommended to divide the class into teams of four or five students. Topics include process descriptions, process safety management, process control, flammability, corrosion, relief device basics, and Design Institute for Emergency Relief Systems (DIERS).

6.10 Summary

SACHE modules provide resources for incorporating process safety into Chemical Engineering courses. The SACHE modules cover the topics of Inherent Safety, Process Hazards, Hazard identification and Risk Assessment, Emergency Relief Systems and fourteen separate case histories.

6.11 References

6.1 Crowl, D. A., Louvar J. F., Chemical Process Safety, Fundamentals with Applications, 3rd Edition, Prentice Hall, Upper Saddle River, NJ: 2011.

6.2 U.S. Chemical Safety and Hazard Investigation Board, Investigation Report, Improving Reactive Hazard Management, Report No. 2001-01-H, October, 2002.
http://www.csb.gov/improving-reactive-hazard-management/

6.3 Essential Practices for Managing Chemical Reactivity Hazard, American Institue of Chemical Engineer, New York, NY, 2003.

7

Process Safety in the Workplace

7.1 What to Expect

This chapter focuses on process safety related duties that engineers can expect during their first year in industry.

7.1.1 Formal Training

Many large companies have formal training matrices not only for new employees, but for employees at many levels and positions throughout their organization. These training matrices enable companies to identify and develop plans to close competency gaps in employees before they are promoted, or change jobs. Table 7.1 is an example of a listing of process safety training for new employees. This is an abbreviated example; a full training matrix will likely include information such as prerequisite course and whether the course is computer based or classroom training. A small or medium sized company may not have a formal training matrix. Each company will have their own method to determine what training a new engineer needs.

An engineer entering industry will discover that the training matrix includes material that was not part of their engineering curricula. Among other things, this may include topics such as occupational safety and health, environmental protection, product safety and stewardship, responsible care management systems, and process safety systems that are unique to the employer. This training may be completed in a classroom setting, computer based training (CBT), or by teleconference or video conference.

Occupational (Personal) Safety: A newly hired engineer can expect to receive in-house orientation and/or training. This training will focus on occupational or personal safety. As described in Section 1.3, the focus of occupational safety is to prevent workers from harm from workplace accidents due to physical and mechanical hazards, such as falls, cuts, repetitive motion injuries, etc. For example, there may be training about Personnel Protective Equipment (PPE) required in various parts of the plant, how to use PPE such as respirators, confined space entry, the hot work permits system, ergonomics and what constitutes an on the job incident or near miss. Occupational safety is an important

Table 7.1 Example simplified process safety training class matrix.

Course	Target Audience	Triggers
Understanding and Managing Flammable Atmospheres	Required for all Engineers, Chemists, involved in design, maintenance and operations	First Two Years
PHA Methodology & Team Leader Training	Recommended for Technical people involved in design, operations, and safety reviews, including MOCs and PHAs. Required for PHA Team Leaders	First Two Years as well as PHA Team Leader Requirement
MOC Safety Review Team Leader Training	Recommended for MOC Core Team Members. Required for MOC Safety Review Team Leaders that have not taken the PHA Team Leader Training Class.	First Two Years
Consequence Assessment	Recommended for people involved in modeling releases of chemicals and energy.	Prior to use of consequence modeling tools.
Pressure Relief Device (PRD) Application	Required for engineers and recommended for designers involved in PRD design, application, sizing and selection.	Prior to involvement in design, application, sizing and selection of PRDs.
Design and Application of SCAI and Safety Instrumented Systems	Required for I&E, Control, and Process Engineers and recommended for Designers involved in Shutdown System design, review, and specification.	Required prior to involvement in Shutdown System review, design or operation OR recommended within the first two years.
Incident Investigation	Recommended for incident investigators and participants	Prior to leading or participating in incident investigations

PROCESS SAFETY IN THE WORKPLACE

part of a chemical facility's safety program. In 2012, 4,628 workers died on the job in the United States.[3] This number demonstrates a need to always maintain a sense of vulnerability. On the other hand, major chemical plants and refineries are far safer than just staying home on a daily basis. Focus on safe operation and past tragic events have made everyone employed in the chemical industry keyed in on process safety practice.

Many organizations are likely to have regular safety meetings and safety seminars covering both occupational and process safety. The new engineer should take advantage of every opportunity to attend these. In some organizations attendance at regular safety meetings is mandatory.

Another feature of many safety programs is a regular walk-around or safety inspection by area personnel. Organizations may have checklists of things to look for during a walk-around that includes items such as:

- are the walkways clear?
- is the lighting adequate?,
- are the transfer hoses worn out?
- are the exits clear or blocked?
- are labels clear or worn out?

Some organizations now have safety critical items listed, and these should be inspected or maintenance records reviewed. Don't be afraid to ask questions or point out something that you think constitutes an unsafe condition. It is part of your job.

In every industrial setting the employer will require that engineers understand and follow all safety rules. Even as a new-hire in industry, engineers are expected to be role models in every aspect of occupational safety, report and/or correct any unsafe condition or act observed. Be prepared to report all incidents, including near-misses, and injuries or potential injuries that occur. While this book is primarily about process safety for the engineer, the importance and attention that should be given to occupational safety and the prevention of injuries to others cannot be stressed enough.

Process Safety: If an organization has a separate process safety department, it is likely that the department will provide the tools and training needed by the first year engineer. Companies without a separate process safety department may offer courses through the safety group and/or contract training out to one or more of the many companies that offer process safety training. They will likely cover many of the topics covered in Table 7.1 and Chapter 6. New engineers should take advantage of such in-house courses if they are offered. Examples of some process

---

[3] Bureau of Labor Statistics, Revisions to the 2012 Census of Fatal Occupational Injuries (CFOI) counts, April 2014, www.bls.gov.

safety tasks that a new engineer may participate in but may not have been taught in school include; the overall structure of the facility safety management system, when and how to conduct PHAs, Management of Change and Pre-Startup Safety Reviews, consequence assessment of releases of hazardous materials, sizing of Emergency Relief Devices, how to do Layer of Protection Analysis, Incident Investigations, how the Safety Lifecycle interfaces with that company's project execution methodology, and the definition of a process safety near-miss and incident in the organization. Each employer will have a different methodology for each topic and it is important that the engineer gain competency in those area that are expected in order to fulfill duties.

If a facility is covered by the US EPA Risk Management Plan, or a Safety Case regulation, there should be a document that provides a summary of the site's process safety program. This is referred as the Risk Management Program Management (RMP) System by the EPA. The facility is required to include a list of personnel with organizational process safety responsibilities, and a list of all RMP documents with their locations. New engineers should familiarize themselves with this document.

If your organization has developed a management of organizational change (MOOC) process in conjunction with the management of change (MOC) process as is described in Chapter 2, the specific training will be detailed through that process. A new-hire engineer should be familiar with the MOOC process and understand expectations set by the person responsible for creating the MOOC document.

7.1.2 Interface with Operators, Craftsmen

Operators are an interface with the process. It has been said that, for any given plant, there are actually three plants – the plant that the engineers and managers think is there, the plant that the operators think is there, and the real plant. Operators know and work with the real plant process every day. New engineers need to make the first plant concept line up with the third plant concept. Operators not only understand how the plant works, but how it can fail. Sometimes things that engineers think are true about the process, procedures, equipment layout are not true.

The same is true for maintenance personnel, electricians and other craftsmen. Such personnel work directly with the process equipment and instrumentation. The Asset Integrity element depends on them. They can provide input to a reliability and/or process engineer on what equipment might be fit or unfit for certain kinds of service.

An anecdote about one engineers experience with an operator in a PHA illustrates an operator's value.

PROCESS SAFETY IN THE WORKPLACE   215

> A HAZOP was being done on an existing process for the first time. The HAZOP team was quite large, including plant engineers, a senior process design engineer, the plant maintenance and reliability engineer, and a maintenance technician. There was extensive discussion about the potential consequences and safeguards during several deviations, with engineers debating the effects of the deviations. Finally the operator said "I don't know about all your logic, but I know that when these events occur, the outlet vent valve opens fully and we get an alarm." During a break the team went to the control room, where the operator deliberately caused the deviations, and in each case, the outlet vent valve opened fully and an alarm went off. It turned out the vent line to a treatment system was a pinch point where the first effect of many deviations announced themselves.

The anecdote illustrates at least two things; first, operators, maintenance personnel and other craftsmen may very well understand the actual cause-effect relationships of process deviations better than the engineers. Second, all people working in the plant, and their opinions should be treated with respect. Their knowledge is important. Avoid a know-it-all attitude and actively listen. If operators and other craftsmen are treated with respect, they will be more willing to tell engineers how a plant really works during a PHA.

As your designs and modifications reach the plant level, you will have to train the operators affected when MOC's, PSSRs, capital and expense projects, require updates to operating procedures and or process safety information.

7.2 New Skills

7.2.1 Non-Technical

The previous paragraph leads to the topic of non-technical skills. Examples of the types of things new engineers must learn include: write concise project status summaries and reports, document process safety information, lead project teams, facilitate meetings, deliver training, participate in (perhaps scribing for or even leading) PHAs such as MOC reviews and/or HAZOPs, incident investigations, and action item management. These require writing skills, public speaking skills as well as human relation skills such as good listening, assertiveness, and respect for other people's opinion.

Hazard Identification and Risk Assessment studies necessarily require identification of scenarios that can lead to impacts such as environmental releases, injuries and fatalities, and property damage. Engineers need to learn to couch these in fact based terms.

For example, a large release of a flammable material inside a congested area can lead to an explosion if ignited (this has been demonstrated in many of the case studies in Chapter 3). One should not write, "This will blow up the entire unit and kill everybody!" To say "if ignited, this can potentially lead to damage to the equipment or processing unit, and one or more fatalities, depending on occupancy levels." is a more precise statement and more neutral. The more quantitative a statement can be made, the better, for example, "a 2 psig overpressure zone extending 200 feet" is preferable to a statement such as "a pretty big explosion".

One area that most every new engineer must master is action item management. Action item management is a critical process safety management skill. Many of the process safety elements described in Chapter 2 generate corrective actions, or action items. For example, a PHA will generate recommendations to reduce risk. MOC and PSSR will generate items to mitigate risk or identify items that must be complete in order to safely start-up a change to process. The purpose of incident investigation is to find and eliminate the causes of process safety incidents. Action items are generated to prevent future process safety incidents. Audits and management reviews often generate lists of actions to correct to prevent process safety incidents.

Action item management can sometimes feel like "busy work" for engineers. However, the new engineer must understand and deal with multiple dynamic priorities. Each action item or recommendation assigned should have the following characteristics:

- A specific person or persons is responsible for resolution
- A specific date is given for resolution
- Completion of an action item or recommendation should be fully documented to include: assumptions, engineering calculations if any, deviation and approval of abnormal closure, approval of extension of closure, a detailed description and proof of that the recommendation is complete, completion of any MOC or PSI (including SOP's and training) associated with closure, and final approval of closure.

A rule of thumb is, if it isn't documented, it isn't done. Each company will have different systems to achieve item resolution, but the elements listed above are usually part of that system. Many companies have an item resolution database in which the engineer must become proficient.

7.2.2 Technical

The previous paragraph leads into the topic of new technical skills. The extent to which a new engineer will need technical skills related to process safety depends on the organization's size and sophistication. These are not necessarily covered in universities. In large organization, these areas are probably handled by Subject Matter Experts (SMEs), or some basic tools will be provided for plant and process engineers, with the SMEs providing advice and using more sophisticated tools when necessary. A special opportunity for new staff is to ask apparently "stupid"

PROCESS SAFETY IN THE WORKPLACE 217

questions of the SMEs, sometimes these have real insight and challenge false assumptions of more senior staff, but do this with respect. In smaller organizations, plant and process engineers may have to do these calculations themselves, or work with contractors specializing in them.

Examples of these tools are Layer of Protection Analysis (LOPA), consequence modeling (see Section 6.6), and Emergency Relief System (ERS) design (see Section 6.7). There are many courses available in these techniques from companies that specialize in these fields. Large organizations may have internal courses. The CCPS publishes many Guideline books that cover these topics.

7.3 Safety Culture

In Section 2.2 the element of process safety culture was described. Process safety culture is the common set of values, behaviors, and norms at all levels in a facility or in the wider organization that affect process safety. The same can apply to safety culture in general, i.e. occupational as well as process safety. The features of a good safety culture are:

- Maintain a sense of vulnerability.
- Rigorously following procedures.
- Empower individuals to successfully fulfill their safety responsibilities.
- Defer to expertise.
- Ensure open and effective communications.
- Establish a questioning/learning environment.
- Foster mutual trust.
- Provide timely response to process safety issues and concerns.

Maintain a sense of vulnerability: Learning and respecting the hazards of a plant was discussed in 7.1.1. Respect for the hazards in a plant usually comes naturally to a new employee. Maintaining respect for the hazards takes some effort. The expression "familiarity breeds contempt" has a basis in truth.

The benefits of the other bullets were described in the anecdote presented in section 7.1.2. Empowerment of individuals, open communication, deference to expertise, mutual trust, and a learning environment were all displayed in the way the operator was willing to present his opinions and how the engineers were willing to listen to him.

A new engineer can get a clue as to a company's safety and process safety culture by asking a few questions or exploring a company's background. A company that is a member of the CCPS, for example, is making a commitment to its process safety program. If a company is not a CCPS member, asking what the process safety program is can reveal a lot about the company. If, for example, the answer is "we comply with regulations" that tells you the company does the minimum it can do.

7.4 Conduct of Operations

Conduct of Operations is a term that describes human factors and tools that are necessary to produce repeatability in performance and consistency in results, and working within the defined operational boundaries. These skills and tools help the manufacturing unit reduce incidents and take the guess work and human error out of operations.

Conduct of operations can be divided into three areas: operating discipline, engineering discipline, and management discipline. A new engineer assigned to interact with a manufacturing unit will have exposure to each area. In this context, the word discipline is used to describe the activities that bring about repeatability and consistency in human interactions with equipment. The CCPS book *Conduct of Operations and Operational Discipline* (Ref. 7.1) covers this topic in more detail.

7.4.1 Operational Discipline

Operational discipline refers to the tools and activities that operators employ to produce consistent results. Even though they are primarily for the operator, unit engineers must be familiar with the content and intent of these activities as to provide feedback and improvement opportunities for each.

Shift or Operating Instructions. Shift instructions are usually produced once per day by an operations specialist or front line supervisor. They are presented in written format either by checklist/form, free-hand notes, or computer generated instructions. The instructions give guidance as to the activities that will occur on shift. Duties of new engineers often include plant tests and trials. How these test and trials proceed include a combination of MOC, Temporary Operating Instructions, Operator Training, and shift notes that serve as temporary operating instructions.

In addition, the new engineer should read the shift instructions each day to become more familiar with the many tasks that are required to keep a manufacturing unit running within specified and approved limits or ranges.

While it is a poor practice to assign inexperienced engineers the responsibility to write shift instructions due to lack of experience, some companies may include this duty, under supervision, as a development exercise.

A good practice would be for the instructions to have:

- A standard format with pre-defined sections
- Specific operating conditions and production targets
- Key performance indicator targets and operating parameter targets

- Any operating constraints
- Special instructions for transient operations such as start-up or shutdown
- Any special process safety, safety, environmental, or reliability considerations
- A means for all operators & other unit personnel such as engineers and process specialist to review the instructions and provide positive acknowledgement that they have reviewed and understand. On return to work these affected personnel should go back and review all instructions missed
- The instructions can be used to communicate any projects, MOC's, special operating instructions, unit tests, special bypass of equipment or safety systems, and any other notable events that will occur in the operating unit.

A good practice is for all support group requests for input to operating instructions be routed through a designated operating instruction person (i.e., there is only one person or position designated to give operating instructions). This includes the new engineer.

Operating shift log or shift notes: A good practice is for each operator to prepare notes or a log each shift and sign it. The log contains a record of all significant events that occurred during the shift, including actions taken and observations made. As with shift instructions, engineers should review each operator's shift notes.

Operators should be trained on documentation expectations for the shift log. The notes are routinely evaluated for completeness through shift supervision's positive verification and performance based audit. Examples of significant events which should be documented are as follows:

- Any process safety concerns or hazardous conditions with latest status and actions taken
- Unscheduled equipment shutdowns, interlock action, SIS or PRD activation. Significant alarm conditions to be noted
- Any environmental problems or concerns with latest status and actions taken
- Details of non-standard operating modes or line-ups
- Status of ongoing maintenance work should be noted, as well as any non-routine contractor activities. (example, open work permits)
- Any problems requiring follow-up or unusual operating occurrences that may require additional investigation (example, developing reliability issues)
- Product quality problems

- All contacts (direct and by phone) with outside parties (neighbors, regulatory agencies, etc.) concerning company business or complaints
- All incidences of interlock bypass or disabling. Any exceedance of a safe operating limit
- Any security issues
- MOC's/ PSSR's completed or begun on shift
- Unit specific operating parameters or KPI's
- Important transfers or production information

Other good practices include:

- Use of a template that covers the above elements is a good practice to insure that each element is covered each shift
- Each operator reviews and provides positive verification of review of all shift notes for their job that have not been previously reviewed
- Unit supervision/ engineers review and provide positive verification of all shift notes and take action regarding issues/ problems noted

Good shift notes detail the problems encountered during the shift and the steps taken to correct. Often unit engineers are called upon to solve problems that arise in the unit. If written thoroughly, the shift notes may give valuable clues that aid the engineer in finding the solutions. Similarly, shift notes provide a platform for proactive problem solving when engineers see subtle changes in comments and notes from operators.

Changes to operator evaluation sheets are often accompanied by an MOC request that is initiated and completed by unit production engineers.

In addition to an engineer's monitoring of operator shift notes, engineers often keep their own shift notes. Research engineers will most certainly keep lab notes and pilot plant observations. The most common application for the unit production engineer is to keep shift notes that detail engineering activities during outages, turnarounds, and during trail or test runs.

Shift Handover. Shift handover is closely related to the shift notes in that the content can be used as a template for discussion. A practice should be in place to ensure the shift handover is thorough and followed and promotes safe and efficient continuance of operations:

- Shift Handover takes place on the job after the oncoming operator is outfitted for work.
- The location is on the job, in an area protected from the environment and excessive unit noise
- Adequate time is given to provide a thorough handover report to the oncoming shift.

- The content elements of the operating shift log are the outline for the handover report.
- Process Leaders discuss all elements of the operating shift log
- Process safety and plant EHS and/or security concerns are emphasized.
- The status of key operational and maintenance activities (including non-routine contractor activities) at shift handover such as equipment start-ups, batch charging, chemical transfers to tanks, railcars and trucks, etc. is identified
- MOC's are communicated
- The bypass log is updated for any safety devices temporarily bypassed or deactivated and what compensating measures are in place.

Communication is always extremely important in a manufacturing unit to ensure continuity of operations. By occasional monitoring of shift handover, the engineer is again afforded a unique opportunity to understand what really goes into operating a chemical plant.

Working Overtime, Weekends, and Holidays. While not explicitly an established topic in operating discipline, working overtime is most definitely a reality for operators and unit production engineers. An engineer entering industry should be aware that they may be called upon to work shift work, 12-hour shifts, straight night shifts, or some other work schedule that is outside of a normal 40 hour straight day job.

Many chemical plants and all refineries operate continuous processes around the clock, each day of the year. Sometimes maintenance is needed on the process as a whole. In this case, the procedures to shut down, make repairs and improvements, and startup safely are carefully planned to optimize safety, minimize risk, and reduce overall cost. These activities necessitate that engineers sometime follow non-standard working schedules. With non-standard work schedules also comes fatigue management. Engineers may be called upon to track operators' individual overtime, and possibly errors, so as to not challenge fatigue rules or guidelines in place in the operating units.

Incident, Event and Near Miss Notification. Expectations for reporting significant events to shift supervision should be made and communicated to operators. Young engineers often take on the role of defining what significant and abnormal events are and train operators on the reporting required by the company. Engineers also monitor that these notifications occur per plan. Operators should be trained on the definition of significant and abnormal events and expectations for communication. Shift logs/notes can be useful in identifying what needs to be reported. There are also regulatory requirements that require reporting of incidents including near misses.

Process Readings and Evaluation. Operators collect information and evaluate that information to determine if predetermined criteria are met. The process parameter readings and evaluation portion of operating discipline and engineering discipline are closely related. Engineers should define what operators need to evaluate, and define how the evaluation is made. Operators or engineers should be thinking about what the process parameter readings mean, not just recording them by rote. When the criteria are not met, the operator must make decisions to take actions to return the process to the predetermined values. For example, if a pump must have a certain discharge pressure to maintain a safe operating limit, that range is defined by the engineer and included on the evaluation sheet. If the range is exceeded, the operator notes the exceedance and notes on the evaluation sheet what is done to return to normal. The engineer (or unit supervision) then closes the loop by shift/ daily review of the evaluation sheets to ensure these concerns are adequately addressed.

One of the most common conduct of operations errors made is to have ineffective process evaluation sheets that have operators simply "taking readings" without evaluating the equipment. When this occurs, actions are not taken on the readings that operators record. Having stated highs and lows on the sheets will help alert operators. At a higher level, SPC monitoring can often pick up deviations before they go "out of range" and cause trouble..

Good practices for operator evaluation should include both outside operator evaluations and board operator evaluations. Since data from the DCS can be printed or stored, it is often assumed that there is no need for an operator to collect or write down that information. However, the purpose of writing the information down isn't to collect it for someone else. The purpose is for the board operator to evaluate that variable or parameter and take pre-determined action if the expected value isn't observed.

Below are a few notes on how to create an effective evaluation sheet along with possible examples of what to evaluate. Engineers need to be familiar with the equipment being evaluated and ensure that pre-defined action is being taken when an abnormal evaluation occurs.

Developing Evaluation Sheet. Evaluation sheets should be designed with engineering and technician input. They should also be reviewed for effectiveness on a periodic basis. When developing evaluation sheets consider the following:

- Map out and document the route taken by the operator
- Each mark that an operator makes on a sheet should be part of the evaluation – either no action is taken, or if outside of a soft or hard evaluation criteria, action is taken
- On return to work, operators review all evaluation sheets for the job they are working that they have not previously reviewed and give positive verification of review
- Unit supervision and engineers review the operator evaluation sheet daily and ensure action is taken for parameters out of range,

PROCESS SAFETY IN THE WORKPLACE

- equipment that needs maintenance or other out of range evaluations noted by the operator
- Each parameter should have the operating range defined and documented, the technical basis, consequence of deviation and pre-defined steps to correct. The operating range and/or expected value should be documented on the evaluation sheet

Specific Equipment Evaluations to Consider:

Safety Equipment

- Safety showers and eyewashes – flush rust out of system and wash dust off the eyewash. Check for adequate flow and proper temperature. Replace dust caps.
- Have maintenance date tags attached so that everyone knows when the previous tests were run.
- Fire protection equipment – fire extinguishers are in place and not out of date, fire monitors are in service with no leaks. Deluge systems are lined-up and not leaking. Evaluate condition of foam systems.
- Breathing air systems – connections are in good condition.
- Radios – spare batteries available and charged.
- LEL meters - calibrated and charged.

Rotating equipment (pumps, motors, fans, blowers, compressors)

- Listen for unusual sounds
- Look for unusual vibrations
- Check for cavitation
- Ensure coupling guards are in place
- Evaluate lubrication levels in sight glasses
- Ensure seal flush flows are adequate
- Look for packing/ seal leaks
- Note unusual smells (some leaks may not be visible)
- Look for oil or grease leaks from bearings
- Check belt and chain condition and that guards are in place
- Look for smoke from rotating parts.

Note, evidence that these types of evaluations have occurred can be noted on an evaluation sheet with a check-mark.

Electrical and Instrument Boxes/ Motor Control Rooms

- Equipment should be clean and dry

- All panel covers are in place, closed and sealed
- Evaluate smoke, smells, and unusual sounds
- Check for adequate lighting
- Areas must be free of trash

Fired Equipment

- Look for gas leaks and/or flames outside of burner box
- Evaluate normality of flame front

Flare operation

- Check seal drum and knock out pot levels
- Evaluate flow to flare
- Evaluate flame/ smoke from flare
- Ensure any required purges are in range

Hazardous Waste storage evaluation

- Inspect for leaks
- Observe dike liner conditions
- Check that dike valves operated closed
- Insure that labels are in place and accurate
- Consider other specific unit considerations

Cooling Towers

- Observe tower for excessive drift
- Evaluate tower for broken louvers, and/or packing
- Look for uniform flow distribution
- Evaluate basin level
- Ensure screens are clear
- Evaluate system for algae & silt buildup
- Observe chemical addition systems for leaks
- Check fans for vibration

Tanks/ Vessels

- Observe tanks for leaks when performing walkarounds
- Review accumulation of liquids in dikes
- Check that water is drained from the roof of any floating roof tanks
- Confirm hatches and strapping ports are closed

- Observe PRDs, especially check for any block valves incorrectly closed isolating the PRD, staining or other evidence that the PRD has lifted, or if there is an intervening bursting disk check that there is no pressure reading on the pressure gauge between it and the PRD. Birds are notorious for building nests in PRD outlets.
- Review pad and vent pressures and line-ups

Heat Exchangers

- Check for head, piping or fitting leaks
- Note unusual or excessive noise or vibration
- Look for changes in normal differential temperatures and pressures
- Back flush exchangers when scheduled or needed.

Process filters and Strainers

- Check for leaks and excessive pressure drop.
- Confirm that bypasses are closed and that the offline filter/strainer is ready to go

Separators and KO Pots

- Monitor levels
- Check for leaks
- Ensure level glasses are clean and readable

Expansion Joints

- Make a general visual inspection ensuring the retaining cable is in place
- Inspect stay bolts
- Evaluate no abnormal growth or bulges, and no leaks

Other types of evaluations to consider with any plant walk through. Look for:

- Process leaks
- Missing plugs, caps, or blinds
- Visible Vapor/ odor
- Leaks / Drips
- Oil sheen/ pools
- Steam leaks
- Noise (e.g. nitrogen or air leaks, cavitating pumps, vibrations, knocking, changes in noise levels)

- Oil level, pressure, temperature, flow
- Unit pressure and temperature gauges
- Unit Flows
- Chemical sewers
- Cooling water
- Control valves
- PRD's
- Suction and discharge blocked
- Telltale gauges
- Frost, sweating
- CAR seals (lock out seals for valves) in place
- Guards in place
- Sprinklers
- Gas detectors
- Emergency shutoff devices
- Unit barricades
- Unit signs (e.g., ingress/ egress)
- Area ingress and egress free
- Safety equipment accessible
- Vehicle access, access to equipment
- Field tags – location and date
- Physical evaluation
- Discoloration of paint and equipment
- Housekeeping
- Sample points/ automatic samples
- Filter inspections, e.g., differential pressure, dP
- Dike valve positions
- LOTO and equipment de-energized or out of service
- Head tanks
- Scheduled pump rotations, back flushes, etc.
- Hoses (condition)
- Area lighting

Safety & Process Safety Evaluation. The operator evaluations are used to collect information and evaluate that information to determine if predetermined criteria are met. This information is documented on an evaluation sheet. A good practice for evaluation is to have a separate sheet for occupational safety and process safety concerns.

Occupational safety evaluations might include inspection of routes of ingress and egress, safety signs, lighting, ladders, hoses, and evaluation of unit safety equipment. Process safety evaluations might also include safety critical equipment or variables that deserve special attention, such as ensuring SIS field bypasses are secured in the safe position. Tell-tale gauges, relief devices, and position critical devices should also be considered. Key process safety alarms should be evaluated.

Defining the appropriate safety and process safety evaluation to be performed is another example of how engineering discipline is closely related to operating discipline. Engineers perform specific reviews with the appropriate process safety information to determine what should be evaluated, and operators complete the inspection. Management discipline then periodically inspects the evaluation sheet to ensure that action is taken on deficiencies noted.

Sample Collection. The process sample schedule should be documented and part of standard operating procedure. Engineers determine this. A good practice is to develop a process map for samples to optimize the round. Well defined sample points are identified and MOC is produced for changes. PPE and other safety precautions are identified for each sample point. The schedule should include:

- Sample point or stream
- Frequency
- Technical basis for the frequency and analysis
- Target ranges for the analysis
- Actions to take in case out of range, including resampling if appropriate

Alarm Disabling & Management. A new engineer may be authorized to approve disabling of alarms. Good practice is to have a written procedure for disabling alarms with the appropriate level of approval. The procedure should also include:

- A means to ensure operators are aware of disabled alarms in their area
- Alternate process indicators to be established and communicated to operators
- Requirements to return disabled alarms to service as quickly as possible. For safety alarms, the compensating measures to use for risk management while the alarm is out of service should be indicated

Position Critical Devices. Engineers are sometimes asked to develop, maintain, and monitor position critical device lists. Position critical devices such as car seals (car seals are devices for physically locking valves in position, see Figure 7.1) can play an important role in the effectiveness of process safety systems. P&IDs should indicate critical car seals as open or closed (CSO or CSC). By itself, a car seal does not necessarily prevent a position critical device from being moved to an unsafe position. It is what the seal represents and how it is managed that keeps the device in the appropriate position, as shown in Figure 7.1.

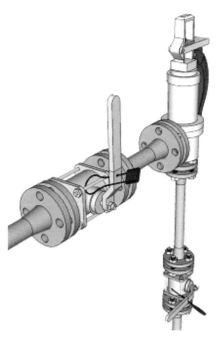

Figure 7.1. Car Seal on a valve handle. Seal can be broken in an emergency if necessary to change the position of a valve, courtesy

Good practices for position critical devices include:

- A written procedure for position critical device management
- Periodic training of the written procedure
- Frequent evaluation checks that the devices are in the proper position
- Performance based audits on the evaluation checks to verify effectiveness of evaluation
- MOC process for removing or changing a position critical device that includes the appropriate level of management review
- PSSR steps that include evaluation of position critical devices.

Emergency Accountability. An aspect of the process safety operational discipline that deserves attention is the emergency accountability system. Engineers are often involved in development and participation of emergency drills and accountability. An important aspect of emergency drills is the effectiveness of the accountability system. Engineers may design and interact with this system.

A good operating discipline practice is to periodically conduct a full scale test of the sites emergency accountability system and correct deficiencies found.

Operator Line-Up. The most fundamental responsibility of an operator is to understand and know the position of every valve in their area of responsibility and to control the energy among all points of material transfer. This responsibility is commonly referred to as "line-up." While engineers usually don't operate or "line-up" equipment, they should clearly understand this aspect of an operator's job and take it into account when designing equipment, development operating procedures, and operator training.

Operators should have a defined responsibility to evaluate all valve positions at the start of each shift. They should be trained on this expectation. Given that the majority of valve positions that an outside operator is responsible for may not change from shift to shift during steady state operation, this is a straight forward task.

It is the non-steady state operation or transient conditions where process safety incidents have a higher probability to occur. Whenever any material is transferred from any point to another the operator must understand and control the transfer of energy. Examples include: opening or closing a valve, starting a pump or compressor, putting any equipment into or out of service, bypassing equipment, and others.

Operators should be trained to "walk-down" the line prior to any and all operational changes listed above. Please see the CCPS website and the "Walk the Line" video online at http://www.aiche.org/ccps/resources/overview/ccps-videos/videos-english/walk-line.

Equipment Labeling. The purpose of proper labeling of equipment and piping is to minimize confusion in plant operations and maintenance activities. Engineers are sometimes called upon to develop and maintain equipment labeling programs that meet these requirements, and other regulatory requirements regarding labeling. Good practices include:

- Major equipment items should be labeled. This includes; tanks, pressure vessels, pumps, compressors, fabricated equipment, control valves and instrumentation
- Spared equipment should be labeled and identified as such. This minimizes chances for confusion.
- Safety instrumented system components
- All piping, including underground entrances and exits
- Utility hose stations
- Safety critical double block and bleed and the intended positions
- Other labels according to plant need

Process Safety Officer. A good practice for conduct of operations is the use of a Process Safety Officer (PSO). A PSO is used to prevent process safety incidents during startups, shutdowns and other critical operating modes by providing an objective independent safety oriented perspective of those activities. Each unit should have a written PSO policy describing qualifications, roles, responsibilities, and a PSO reference manual.

Those responsible for unit operations should:

- Identify and document safety critical startups, shutdowns, and other operating modes which require a PSO
- Document qualifications of the PSO
- Schedule PSO
- Continuous improvement of PSO effectiveness

PSO Responsibilities

- Be present during safety critical startup, shutdowns, and other operating modes which require a PSO
- Have no other duties that distract from their PSO responsibilities
- Has authority to stop any activity that is unsafe

PSO Qualifications

- Have an in-depth knowledge of the hazards of the process
- Have a working knowledge of unit operations
- Understand all applicable standard operating procedures

Included in the PSO Reference Manual:

- Process safety incident reports from PHA
- Roles and responsibilities
- Hazards of unit from PHA
- Training material for the PSO

In those companies that have a PSO, more experienced engineers may be called upon to fulfill this or similar roles.

7.4.2 Engineering Discipline

The second broad area of conduct of operations is called engineering discipline. While Operating Discipline is concerned with operating activities to achieve repeatability in results, Engineering Discipline is concerned with how engineers monitor both the operators' activities and the technical information in an operating unit in order to produce consistency in operations and repeatability in results.

Key Performance Indicators. Engineers should identify key performance indicators (KPI's) used to track process safety, environmental, reliability, and economic optimization and have a defined schedule of review and action taken. Documentation includes:

- Operating Range
- Technical basis for the operating range
- Steps to correct
- Who is responsible to monitor (and backup)
- Frequency to monitor

These KPI's are reviewed on a daily basis and action taken for out of range KPI's.

Safe Operating Limits. Safe operating limits, also known as the safe operating envelope, are a subset of key performance indicators listed above. They are normally set for parameters such as temperature, pressure, level flow or concentration based on the process safety information of the process operation outside of which might cause a negative or undesirable consequence. These limits should be defined and documented. They become process safety information and are used or referenced in: PHA, MOC, PSSR, SOP's, and operator certification/ recertification. Engineers should monitor that this important piece of PSI is accurate and used in these process safety management systems.

Accountability. Engineers and unit supervision provide a key check point to insure conduct of operation management systems are in place and working as intended. There should be an accountability check by engineers and unit supervision to prevent drift and take action on findings. Good engineering discipline practices include:

Management Practice

- Site management periodically completes a performance audit on the various operating elements of conduct of operations
- Conduct of operations is an element in process safety management review. Performance based audits are reviewed and action taken to correct deficiencies

Unit Supervision/ Unit Engineers

- Daily review of shift instructions for accuracy and ensure that instructions are followed. Provide positive verification of review (sign off on instructions.)
- Daily review of operator shift notes for completeness and accuracy. Take action on deficiencies noted. Provide positive verification of review (sign off that notes were reviewed.)

- Daily review of operator evaluation sheets. Take action on equipment that is noted out of service or in need of repair and of equipment operated outside of the defined operating range.
- Look for readings outside of the range with no comment. Take action to understand why no comment appears and correct if necessary.
- Look for "pencil whipped" readings – the same number each evaluation where there should be variability. Take action by asking why and correct if necessary.

7.4.3 Management Discipline

The last broad area of conduct of operations and perhaps the most important is called Management Discipline. The case studies in Chapter 3 show that there are almost always several management system causes underlying a process safety incident. Management plays a key role in reducing human error by providing discipline in conduct of operations.

Management discipline is defined as the actions that management takes to ensure that repeatable results are obtained. Developing a process safety culture begins with leadership, and leadership begins with developing systems that clearly lay out activities that employees perform to achieve results. A few good management system practices that enhance conduct of operations follow. This list is not all inclusive but rather is presented to illustrate a few areas where management discipline should be employed to achieve consistent results. Engineers often have tasks and duties that support these management discipline activities and indeed, when more experienced is gained, may be leading the management discipline topics.

Daily Review of Operations. Many companies have daily production reviews with staff to define the current manufacturing issues and plans to address. A more important aspect of the daily meeting is a review of all incidents to insure proper classification, investigation, and resources devoted for prevention of a repeat. Correction of near miss causes prevents process safety incidents.

The engineering staff and site director should review each process safety incident report. The purpose is to ensure proper classification and that appropriate investigation is completed in a timely manner. Daily review provides a good tool for the site director to communicate directly to section leaders on what's important to work on & evaluate effectiveness of closure. Good practices include:

- Management encourages timely reporting of all incidents and near misses
- Site director and direct reports review all incidents each Monday through Friday and ensures each is classified correctly.
- Establishing a category of near miss incidents called High Potential incidents (HiPo's), that need special attention as these could have been very serious in other circumstances.

- All incidents are investigated. The review team determines the rigor, methodology, and team members for investigation teams; produces investigation team charters with deliverables and time lines; follow-up on open investigations; and follow-up on effective closure of action items.
- A system to identify repeat incidents is in place for the daily review.
- All employees review all incidents to raise awareness of near misses and incidents.
- Management supports recommendations and follows through with the resources to minimize the incident or near miss from occurring again.

Site Procedures & Management Systems. A well written corporate procedure should provide clear unambiguous guidance that tells sites what to do. A well written site procedure should provide equally clear unambiguous guidance that describes who does what in order to fulfill the requirements of the corporate procedure. Engineers are often asked to write or update site procedures and therefore must be very familiar with the site procedures and management systems. Some things to consider:

- Consider a separate site procedure that assigns process safety responsibilities and accountabilities to specific individuals. Ensure that the plan is updated as part of the sites management of organizational change procedure
- In site procedures, include a well thought out section that clearly defines tasks and responsibilities for each document and process safety management system. There should be enough granularity in the definition of responsibility that it is clear to the reader which position(s) are responsible for what
- Review site procedure responsibility section annually
- Provide annual training on site procedure documents to insure all positions understand their role
- Periodically dialog with direct reports to ensure that they understand their role as defined in the site procedures.

Housekeeping. Poor housekeeping is a warning sign of an ineffective safety program. Providing an orderly, clean worksite improves performance and morale, and it reduces injuries and process safety incidents. Provide clear expectation for all personnel regarding housekeeping and inspect performance daily or on a shift basis. Performance will follow. Engineers often participate in housekeeping inspections and sometimes are asked to lead housekeeping improvement efforts.

Administrative Practices

- Develop a written housekeeping procedure that assigns specific duties and expectations
- Monitor and assess effectiveness of the housekeeping program on a shift to shift basis.
- Immediately correct deficiencies

Office Housekeeping

- All offices should be kept clear of debris.
- Spend a little money on good office furniture and provide adequate storage
- Do not stack books and papers on desks and credenzas
- Do not allow graffiti in offices, restrooms, or operating units. Correct these deficiencies on the shift on which they are identified.

Unit Housekeeping

- Assign specific responsibilities. Each employee should participate in some type of housekeeping activity each shift or day of work
- Unit supervision must evaluate each employee's housekeeping effort each shift and correct deficiencies as they are found

Safe Work Permits. Engineers are sometimes asked to write permits, or audit the permitting process and should therefore be familiar with the various types of permits that exist in industry such as: safe work, hot work, excavation, elevation, confined space entry, and more. These topics are usually termed life-critical, because they are exactly that. There will be a detailed procedure for a new engineer to become familiar with: only designated people can issue a permit, only designated people can receive a permit. Usually a multi-part form is used so that copies can be present in the control room, the field where the work is executed, and in the record keeping area. The permit form will identify a range of safety actions that must be complied with (e.g. flammable or toxic gas test, the frequency for repeating such tests, the need for special PPE, whether an extinguisher is to be at the worksite, etc.). Also it is now common practice to require a Job Security Assessment (JSA) for the whole team to review before starting work. This is a documented system. Each requirement must be clearly and unambiguously understood by all that participate in the activity and each element must be followed at all times, or fatalities could occur. Part of the sign-off process for work permits is evaluating the job in the field at the end of each shift, regardless if work is carried over or incomplete. The operator must ensure that maintenance or contractors will leave the worksite in an acceptably clean manner.

Fatigue Management. Several major process safety incidents across the globe have contributing causes relating to fatigue management. The fact is that tired and overworked employees make more errors. This is especially true for major project work such as turnarounds during the most critical phase of operation; startup following a long working period. Yet, many companies still do not have formal fatigue management procedures. A good reference in developing a fatigue management procedure is found in API RP 755, Fatigue Risk Management Systems for Personnel in the Refining and Petrochemical Industries [Ref. 7.3]. Other good practices that engineers may be involved with include:

- Measure overtime by site, unit, shift, and person
- Set overtime targets and manage to meet them
- Develop a written site policy for fatigue management
- Develop a policy that limits the total number of consecutive shifts that can be worked and mandate a minimum rest time between shift
- Develop a separate special projects & turnaround policy
- Allow for deviation to policy with appropriate level of formal management approval

Restrict Access to Console Operators/ Minimize Distraction. A board operator can have literally thousands of control points to monitor in a process, each with various attributes. The mode, set point, process variable and alarm status should be known at all times. Unnecessary distractions divert the board operators attention from their primary role; safe operation of the manufacturing unit.

Implementing these best practices can have a large effect in reducing human error for our console operators.

- Eliminate all non-work related materials from the control room including but not limited to: radios, books, newspapers, iPods, and iPads. Restrict the use of personal phones and text messaging devices
- Restrict access to the board operator to those who must have access to the control room only. e.g., the unit engineer, shift leader, process specialist, and in some cases, outside operators
- The control room should not be a meeting or resting place for outside operators, contractors, and other plant employees
- Board operators should not be the focal point for issuing maintenance and contractor permits, unless they are responsible to issue the permit
- Avoid meetings in the control room with an on shift board operator.

Communications. Process safety goals, objectives, site metrics, and action plans should be clearly and frequently communicated to site employees so that everyone understands where performance is with regard to process safety and where the site is heading. Management should monitor these types of communications to achieve performance. Engineers often contribute to or are in charge of various types of communications. These include:

Communication Boards

- Assign responsibility for periodically monitoring all bulletin boards across the site including control rooms, office buildings, meeting rooms, etc. to insure that each board has a purpose, information displayed professionally and is kept up to date
- Especially pay attention to control rooms, and remove all non-professional wall materials, sticky notes, papers, and other distractive materials
- Define process safety communication bulletin boards in key locations including control rooms and meeting rooms and display: site process safety goals, objectives, metrics/graphs/trends, any action plans and current status and any other special process safety communications

Other Written Communication

- Post or distribute the CCPS Process Safety Beacon newsletters for use at tailgate meetings to focus attention on process safety
- Consider site newsletters and periodic emails that draw attention to site metrics and objectives

Verbal Communication

- Consider a dedicated day for process safety topics at tailgates
- Have a separate process safety section in the periodic sequential safety meetings that have detailed topics. Include metric and objectives review

Continuity of Operations. Continuity of operations is an attempt to provide a sense of ownership of the equipment by operations by minimizing interruptions between jobs, maintenance and capital work, and ensuring effective personal contact among unit supervision and operators.

- Ensure appropriate overlap in shift relief where possible, between supervision and shift workers
- Where possible, do not rotate shift workers out of a job within any shift rotation cycle. That is, operators often rotate among the various unit jobs: outside work, sample operator, and board operator to name a few common functions. Rotation among these positions should not occur in

the same rotating shift cycle. For example, work the board job during the day cycle and rotate to the outside job during the next night cycle in order to provide continuity of operation.

Accountability. Part of management discipline is to ensure that a defined list of expectations exist for front line supervisors, operators, and crafts. Train employees on those expectations and hold them accountable for proper execution. This is the very definition of management. See the section above on engineering discipline accountability for ideas on how to verify performance.

7.4.4 Other Conduct of Operations Topics for the New Engineer

Observation and attention to detail: New engineers need to develop this skill. There is no easy way to do this. Safety walk-arounds, described in Section 7.1.1, provide an opportunity to develop these skills, as are tasks such as review of operator logs, checking and updating Piping and Instrumentation Diagrams (P&IDs).

A learning attitude: Education does not end with a diploma. In section 4.7, resources for further learning about process safety were listed that new engineers can take advantage of. New engineers should also look for opportunities to learn more through company seminars, if they are offered. Inform others who don't know of the existence of these resources. Make an effort to attend local section AIChE meetings. Make an effort to attend at least one national conference annually in your field of engineering. The AICHE GCPS meeting held annually in the Spring is a wealth of present practice in process safety by many industries worldwide.

Listening to operators, see Section 7.1.2, is a way to develop a learning attitude. Becoming familiar with the Process Safety Information, recent incident reports and documentation such as the RMP, or Safety Case, if available, is another way to accomplish this.

Recognize hazards: This was discussed in section 7.1.1, and in the need to learn what near misses and incidents are (Section 4.6). The lessons from incidents described in Chapter 3 should be reviewed on a regular basis. New engineers should look for ways that the lessons from those incidents can be applied to their plant. Once again, the safety walk-around is another instrument for learning to recognize hazards in the workplace.

Self-check and peer-check: As with observation and attention to detail, this is a skill that must be developed. Opportunities to develop these skills occur when doing and documenting engineering calculations.

Standards of conduct: Engineers must follow the rules learned during the initial training period as an example to operators. An engineer is almost always viewed as a leader to someone. If, for example, an engineer does not wear all the required PPE in an area a bad example can be set, which makes it more difficult to enforce the PPE guidelines in a plant (a desired outcome would be if the operators remind the engineers when they are not following rules and procedures). Similarly, taking short-cuts, such as allowing an operator to bypass a procedure, also sets a bad example for future actions.

7.5 Summary

The first year is a massive learning period for a new engineer. Initial learning, discussed in Section 7.1, can take place through formal training and listening to operators about how processes really work. Larger companies are moving towards developing training matrices for not only new engineers, but all levels of employees. Ongoing learning resources were provided in Chapter 4.7.

The new engineer will also have to develop many non-technical skills, such as, clear and concise writing, speaking, time management, and action-item management. Some organizations will offer training in-house or bring in outside trainers to do this. In smaller companies an engineer may have to seek out such training.

The safety culture is the common set of values and beliefs of the organization. A new engineer will be able to detect what kind of safety culture has by observing the attitudes of people; do they follow the rules and procedures or take short cuts, know and respect the hazards of the process or seem oblivious to them, communicate well between functions or do not each other? The new engineer should adopt good safety culture /conduct of operation practices as outlined in this chapter.

Conduct of operations includes operational, engineering and management discipline. Elements of operational discipline were covered in some detail to provide a new engineer with a sense of what to expect when starting in industry.

Enjoy your chosen profession. Remember the first tenet in the AIChE Code of Ethics: "Hold paramount the safety, health and welfare of the public and protect the environment in performance of their professional duties."[4]

7.6 References

7.1 Conduct of Operations and Operational Discipline, Center for Chemical Process Safety, American Institute of Chemical Engineers, New York, New York, 2011.

7.2 Process Safety Leading and Lagging Metrics, Center for Chemical Process Safety, American Institute of Chemical Engineers, New York, New York, Revised 2011. (http://www.aiche.org/sites/default/files/docs/pages/metrics%20english%20updated.pdf)

---

[4] http://www.aiche.org/about/code-ethics

7.3 API RP 754, Process Safety Performance Indicators for the Refining & Petrochemical Industries, American Petroleum Institute, 1st Ed., Washington, DC, 2010.

7.4 API RP 755, Fatigue Risk Management Systems for Personnel in the Refining and Petrochemical Industries, American Petroleum Institute, 1st Ed., Washington, DC., 2010.

APPENDIX A – EXAMPLE RAGAGEP LIST

Table A.1 is a partial Recognized and Generally Accepted Good Engineering Practice (RAGAGEP) list for a fictitious chemical manufacturer, XYZ Chemicals. It is provided as an example only and not meant to be a complete list. These are external standards. Internal best practices developed by an organization may be used as well, as long as they are not less stringent than an existing external standard.

Recognized and generally accepted good engineering practice is a term originally used by OSHA, stemming from the selection and application of appropriate engineering, operating, and maintenance knowledge when designing, operating and maintaining chemical facilities with the purpose of ensuring safety and preventing process safety incidents.

RAGAGEP involves the application of engineering, operating or maintenance activities derived from engineering knowledge and industry experience based upon the evaluation and analyses of appropriate internal and external standards, applicable codes, technical reports, guidance, or recommended practices or documents of a similar nature. RAGAGEP can be derived from singular or multiple sources and will vary based upon individual facility processes, materials, service, and other engineering considerations.

Table A-1. RAGAGEP List for XYZ Chemicals.

Topic	Code
Atmospheric Tanks	API 620: Design and Construction of Large, Welded, Low-pressure Storage Tanks
Chemical Specific Codes	
Chlorine	Chlorine Institute Pamphlet 5) Bulk Storage of Liquid Chlorine
	Chlorine Institute Pamphlet 6) Piping Systems for Dry Chlorine
	Chlorine Institute Pamphlet 9) Chlorine Vaporing Systems
Peroxides	NFPA 430: Code For the Storage Of Liquid and Solid Oxidizers
Compressed Gases	Compressed Gas Association P-22: The Responsible Management and Disposition of Compressed Gases and their Cylinders

Table A-1. RAGAGEP List for XYZ Chemicals, continued.

Topic	Code
Fired Equipment	NFPA 85: Boiler and Combustion Systems Hazards Code
	NFPA 86: Standard For Ovens And Furnaces
	FM 6-0: Industrial Heating Equipment, General
	FM 6-9: Industrial Ovens and Dryers
	FM 6-10: Process Furnaces
	FM 7-99: Hot Oil Heaters
	API 521: Pressure-Relieving and Depressuring Systems
	API 537: Flare Details For General Refinery and Petrochemical Service
Flammable Liquids	NFPA 30: Flammable and Combustible Liquids Code
	NFPA 77: Recommended Practice on Static Electricity
Heat Exchangers	TEMA: Standards of the Tubular Exchanger Manufacturers Association
	API 510: Pressure Vessel Inspection Code: In-Service Inspection, Rating, Repair, and Alteration
Instrumentation and Controls	ISA-18.2 Management of Alarm Systems for the Process Industries
	ISA-84.91.01 Identification and Mechanical Integrity of Safety Controls, Alarms, and Interlocks in the Process Industry
	ISA-84.00 Functional Safety: Safety Instrumented Systems for the Process Industry Sector
	ISA-101 (Draft) Human Machine Interfaces for Process Automation Systems
Plant Buildings	API 752: Management of Hazards Associated With Location of Process Plant Permanent Buildings
	API 753: Management of Hazards Associated With Location of Process Plant Portable Buildings
Pressure Vessels	ASME Section VIII – Pressure Vessels
	API 510: Pressure Vessel Inspection Code: In-Service Inspection, Rating, Repair, and Alteration

APPENDIX A

Table A-1. RAGAGEP List for XYZ Chemicals, continued.

Topic	Code
Solids Handling Equipment	NFPA 654: Standard for the Prevention of Fires and Dust Explosions from the Manufacturing, Processing, and Handling of Combustible Particulate Solids NFPA 68: Standard on Explosion Protection by Deflagration Venting NFPA 69: Standard on Explosion Prevention Systems FM 7-76: Prevention and Mitigation of Combustible Dust Hazards

APPENDIX B – LIST OF CSB VIDEOS

The U.S. Chemical Safety Board (CSB) creates videos of many incidents. Table B.1 lists a series of CSB reports and their associated videos. As of the publication of this book, these videos are online at www.csb.gov/videos/. Reports can also be searched online at www.csb.gov/investigations. Each video can be accessed from its investigation report website.

The incidents in Table B.1 are categorized by topic: Asset Integrity and Reliability, Combustible Dusts, Laboratory Hazards, Reactive Chemicals, and Safe Work Permits. Several incidents could fall under multiple topics, and are grouped under "other". Some CSB videos are not directly related to incidents, such as safety messages. Those videos are not listed here.

Table B.1 List of CSB Videos

Investigation	Year of Incident	Video
Asset Integrity and Reliability		
NDK Crystal Inc. Explosion with Offsite Fatality	2009	Falling Through the Cracks
Silver Eagle Refinery Flash Fire and Explosion and Catastrophic Pipe Explosion	2009	Silver Eagle Refinery Explosion Surveillance Footage
Tesoro Refinery Fatal Explosion and Fire	2010	Animation of Explosion at Tesoro's Anacortes Behind the Curve The Human Cost of Gasoline
DuPont Corporation Toxic Chemical Releases	2010	Fatal Exposure: Tragedy at DuPont Animation of January 23, 2010 Phosgene Accident
Chevron Refinery Fire	2012	Chevron Richmond Refinery Fire Animation Surveillance Video from the August 6 Accident at the Chevron Refinery in Richmond, CA
Freedom Industries Chemical Release	2014	Freedom Industries Tank Dismantling
Combustible Dusts		
Imperial Sugar Company Dust Explosion and Fire	2008	Inferno: Dust Explosion at Imperial Sugar
AL Solutions Fatal Dust Explosion	2010	Combustible Dust: Solutions Delayed Combustible Dust: An Insidious Hazard
Hoeganaes Corporation Fatal Flash Fires	2011	Iron in the Fire Dust Testing

Table B.1 List of CSB Videos, continued.

Investigation	Year of Incident	Video
Laboratory Hazards		
Texas Tech University Chemistry Lab Explosion	2010	Experimenting with Danger
Key Lessons for Preventing Incidents from Flammable Chemicals in Educational Demonstrations	2014	After the Rainbow
Reactive Chemicals		
Preventing Harm from NaHS		Preventing Harm from NaHS
Improving Reactive Hazard Management	2000	
BP Amoco Thermal Decomposition Incident	2001	Reactive Hazards
Synthron Chemical Explosion	2006	
Formosa Plastics Vinyl Chloride Explosion	2004	Explosion at Formosa Plastics (Illinois)
CAI / Arnel Chemical Plant Explosion	2006	Blast Wave in Danvers
T2 Laboratories Inc. Reactive Chemical Explosion,	2007	Runaway: Explosion at T2 Laboratories
Bayer CropScience Pesticide Waste Tank Explosion	2008	Fire in the Valley Inherently Safer: The Future of Risk Reduction
West Fertilizer Explosion and Fire (Investigation ongoing as of publishing of the book) (Ammonium Nitrate)	2013	CSB Video Documenting the Blast Damage in West, Texas
Safe Work Permits		
Hazards of Nitrogen Asphyxiation	2005	
Final Report: Power Point Presentation on Nitrogen Hazards	2003	Hazards of Nitrogen Asphyxiation
Final Report: Safety Bulletin - Hazards of Nitrogen Asphyxiation	2003	
Valero Refinery Asphyxiation Incident	2005	
Partridge Raleigh Oilfield Explosion and Fire	2006	Death in the Oilfield
E. I. DuPont De Nemours Co. Fatal Hotwork Explosion	2010	Hot Work: Hidden Hazards
Packaging Corporation Storage Tank Explosion Partridge Raleigh Oilfield Explosion and Fire Bethune Point Wastewater Plant Explosion, 2006 Motiva Enterprises Sulfuric Acid Tank Explosion Seven Key Lessons to Prevent Worker Deaths During Hot Work In and Around Tanks		Dangers of Hot Work

APPENDIX B

Table B.1 List of CSB Videos, continued.

Investigation	Year of Incident	Video
Xcel Energy Company Hydroelectric Tunnel Fire	2007	No Escape: Dangers of Confined Spaces
Bethune Point Wastewater Plant Explosion	2006	Public Worker Safety
Formosa Plastics Propylene Explosion	2005	Fire at Formosa Plastics (Texas)
Praxair Flammable Gas Cylinder Fire	2005	Dangers of Propylene Cylinders
Acetylene Service Company Gas Explosion	2005	Dangers of Flammable Gas Accumulation
Other		
DPC Enterprises Festus Chlorine Release	2002	Emergency Preparedness: Findings from CSB Accident Investigations
Sterigenics Ethylene Oxide Explosion	2004	Ethylene Oxide Explosion at Sterigenics
BP America Refinery Explosion	2005	Anatomy of a Disaster
EQ Hazardous Waste Plant Explosions and Fire	2006	Emergency in Apex
Valero Refinery Propane Fire	2007	Fire from Ice
Barton Solvents Explosions and Fire	2007	Static Sparks Explosion in Kansas
Little General Store Propane Explosion	2007	Half An Hour to Tragedy
CITGO Refinery Hydrofluoric Acid Release and Fire	2009	Surveillance video from July 19, 2009, fire and explosion at the CITGO Corpus Christi Refinery
ConAgra Natural Gas Explosion and Ammonia Release	2009	Deadly Practices
Kleen Energy Natural Gas Explosion	2010	
Macondo Well Blowout	2010	Deepwater Horizon Blowout Animation
Donaldson Enterprises, Inc. Fatal Fireworks Disassembly Explosion and Fire	2011	Deadly Contract
Millard Refrigerated and Ammonia Release	2015	Shock to the System

APPENDIX C – REACTIVE CHEMICALS CHECKLIST

This checklist is adapted from a CCPS Safety Alert; *A Checklist for Inherently Safer Chemical Reaction Process Design and Operation*, March 1, 2004. Copyright 2004.

C.1 Chemical Reaction Hazard Identification

1. Know the heat of reaction for the intended and other potential chemical reactions.

There are a number of techniques for measuring or estimating heat of reaction, including various calorimeters, plant heat and energy balances for processes already in operation, analogy with similar chemistry (confirmed by a chemist who is familiar with the chemistry), literature resources, supplier contacts, and thermodynamic estimation techniques. You should identify all potential reactions that could occur in the reaction mixture and understand the heat of reaction of these reactions.

2. Calculate the maximum adiabatic temperature for the reaction mixture.

Use the measured or estimated heat of reaction, assume no heat removal, and that 100% of the reactants actually react. Compare this temperature to the boiling point of the reaction mixture. If the maximum adiabatic reaction temperature exceeds the reaction mixture boiling point, the reaction is capable of generating pressure in a closed vessel and you will have to evaluate safeguards to prevent uncontrolled reaction and consider the need for emergency pressure relief systems.

3. Determine the stability of all individual components of the reaction mixture at the maximum adiabatic reaction temperature.

This might be done through literature searching, supplier contacts, or experimentation. Note that this does not ensure the stability of the reaction mixture because it does not account for any reaction among components, or decomposition promoted by combinations of components. It will tell you if any of the individual components of the reaction mixture can decompose at temperatures which are theoretically attainable. If any components can decompose at the maximum adiabatic reaction temperature, you will have to understand the nature of this decomposition and evaluate the need for safeguards including emergency pressure relief systems.

4. Understand the stability of the reaction mixture at the maximum adiabatic reaction temperature.

Are there any chemical reactions, other than the intended reaction, which CCPS Safety Alert, March 1, 2004 3 can occur at the maximum adiabatic reaction temperature? Consider possible decomposition reactions, particularly those which generate gaseous products. These are a particular concern because a small mass of reacting condensed liquid can generate a very large volume of gas from the reaction products, resulting in rapid pressure generation in a closed vessel. Again, if this is possible, you will have to understand how these reactions will impact the need for safeguards, including emergency pressure relief systems. Understanding the stability of a mixture of components may require laboratory testing.

5. Determine the heat addition and heat removal capabilities of the pilot plant or production reactor.

Don't forget to consider the reactor agitator as a source of energy – about 2550 Btu/hour/horsepower. Understand the impact of variation in conditions on heat transfer capability. Consider factors such as reactor fill level, agitation, fouling of internal and external heat transfer surfaces, variation in the temperature of heating and cooling media, variation in flow rate of heating and cooling fluids.

6. Identify potential reaction contaminants.

In particular, consider possible contaminants which are ubiquitous in a plant environment, such as air, water, rust, oil and grease. Think about possible catalytic effects of trace metal ions such as sodium, calcium, and others commonly present in process water. These may also be left behind from cleaning operations such as cleaning equipment with aqueous sodium hydroxide. Determine if these materials will catalyze any decomposition or other reactions, either at normal conditions or at the maximum adiabatic reaction temperature.

7. Consider the impact of possible deviations from intended reactant charges and operating conditions.

For example, is a double charge of one of the reactants a possible deviation, and, if so, what is the impact? This kind of deviation might affect the chemistry which occurs in the reactor – for example, the excess material charged may react with the product of the intended reaction or with a reaction solvent. The resulting unanticipated chemical reactions could be energetic, generate gases, or produce unstable products. Consider the impact of loss of cooling, agitation, and temperature control, insufficient solvent or fluidizing media, and reverse flow into feed piping or storage tanks.

APPENDIX C

8. Identify all heat sources connected to the reaction vessel and determine their maximum temperature.

Assume all control systems on the reactor heating systems fail to the maximum temperature. If this temperature is higher than the maximum adiabatic reaction temperature, review the stability and reactivity information with respect to the maximum temperature to which the reactor contents could be heated by the vessel heat sources.

9. Determine the minimum temperature to which the reactor cooling sources could cool the reaction mixture.

Consider potential hazards resulting from too much cooling, such as freezing of reaction mixture components, fouling of heat transfer surfaces, increases in reaction mixture viscosity reducing mixing and heat transfer, precipitation of dissolved solids from the reaction mixture, and a reduced rate of reaction resulting in a hazardous accumulation of unreacted material.

10. Consider the impact of higher temperature gradients in plant scale equipment compared to a laboratory or pilot plant reactor.

Agitation is almost certain to be less effective in a plant reactor, and the temperature of the reaction mixture near heat transfer surfaces may be higher (for systems being heated) or lower (for systems being cooled) than the bulk mixture temperature. For exothermic reactions, the temperature may also be higher near the point of introduction of reactants because of poor mixing and localized reaction at the point of reactant contact. The location of the reactor temperature sensor relative to the agitator, and to heating and cooling surfaces may impact its ability to provide good information about the actual average reactor temperature. These problems will be more severe for very viscous systems, or if the reaction mixture includes solids which can foul temperature measurement devices or heat transfer surfaces. Either a local high temperature or a local low temperature could cause a problem. A high temperature, for example, near a heating surface, could result in a different chemical reaction or decomposition at the higher temperature. A low temperature near a cooling coil could result in slower reaction and a buildup of unreacted material, increasing the potential chemical energy of reaction available in the reactor. If this material is subsequently reacted because of an increase in temperature or other change in reactor conditions, there is a possibility of an uncontrolled reaction due to the unexpectedly high quantity of unreacted material available.

11. Understand the rate of all chemical reactions.

It is not necessary to develop complete kinetic models with rate constants and other details, but you should understand how fast reactants are consumed and generally how the rate of reaction increases with temperature. Thermal hazard calorimetry testing can provide useful kinetic data.

12. Consider possible vapor phase reactions.

These might include combustion reactions, other vapor phase reactions such as the reaction of organic vapors with a chlorine atmosphere, and vapor phase decomposition of materials such as ethylene oxide or organic peroxide.

13. Understand the hazards of the products of both intended and unintended reactions.

For example, does the intended reaction, or a possible unintended reaction, form viscous materials, solids, gases, corrosive products, highly toxic products, or materials which will swell or degrade gaskets, pipe linings, or other polymer components of a system? If you find an unexpected material in reaction equipment, determine what it is and what impact it might have on system hazards. For example, in an oxidation reactor, solids were known to be present, but nobody knew what they were. It turned out that the solids were pyrophoric, and they caused a fire in the reactor.

14. Consider doing a Chemical Interaction Matrix and/or a Chemistry Hazard Analysis.

These techniques can be applied at any stage in the process life cycle, from early research through an operating plant6. They are intended to provide a systematic method to identify chemical interaction hazards and hazards resulting from deviations from intended operating conditions.

C.2 Reaction Process Design Considerations

1. Rapid reactions are desirable.

In general, you want chemical reactions to occur immediately when the reactants come into contact. The reactants are immediately consumed and the reaction energy quickly released, allowing you to control the reaction by controlling the contact of the reactants. However, you must be certain that the reactor is capable of removing all of the heat and any gaseous products generated by the reaction.

2. Avoid batch processes in which all of the potential chemical energy is present in the system at the start of the reaction step.

If you operate this type of process, know the heat of reaction and be confident that the maximum adiabatic temperature and pressure are within the design capabilities of the reactor.

3. Use gradual addition or "semi-batch" processes for exothermic reactions.

The inherently safer way to operate exothermic reaction process is to determine a temperature at which the reaction occurs very rapidly. Operate the reaction at this temperature, and feed at least one of the reactants gradually to limit the potential energy contained in the reactor. This type of gradual addition process is often called "semi-batch." A physical limit to the possible rate of addition of the limiting reactant is desirable – a metering pump, flow limited by using a small feed line, or a restriction orifice, for example. Ideally, the limiting reactant should react immediately, or very quickly, when it is charged. The reactant feed can be stopped if necessary if there is any kind of a failure (for example, loss of cooling, power failure, loss of agitation) and the reactor will contain little or no potential chemical energy from unreacted material. Some way to confirm actual reaction of the limiting reagent is also desirable. A direct measurement is best, but indirect methods such as monitoring of the demand for cooling from an exothermic batch reactor can also be effective.

4. Avoid using control of reaction mixture temperature as the only means for limiting the reaction rate.

If the reaction produces a large amount of heat, this control philosophy is unstable – an increase in temperature will result in faster reaction and even more heat being released, causing a further increase in temperature and more rapid heat release. If there is a large amount of potential chemical energy from reactive materials, a runaway reaction occurs. This type of process is vulnerable to mechanical failure or operating error. A false indication of reactor temperature can lead to a higher than expected reaction temperature and possible runaway because all of the potential chemical energy of reaction is available in the reactor. Many other single failures could lead to a similar consequence – a leaking valve on the heating system, operator error in controlling reactor temperature, failure of software or hardware in a computer control system.

5. Account for the impact of vessel size on heat generation and heat removal capabilities of a reactor.

Remember that the heat generated by a reactive system will increase more rapidly than the capability of the system to remove heat when the process is operated in a larger vessel. Heat generation increases with the volume of the system – by the cube of the linear dimension. Heat removal capability increases with the surface area of the system, because heat is generally only removed through an external surface of the reactor. Heat removal capability increases with the square of the linear dimension. A large reactor is effectively adiabatic (zero heat removal) over the short time scale (a few minutes) in which a runaway reaction can occur. Heat removal in a small laboratory reactor is very efficient, even heat leakage to the surroundings can be significant. If the reaction temperature is easily controlled in the laboratory, this does not mean that the temperature can be controlled in a plant scale reactor. You need to obtain the heat of reaction data discussed previously to confirm that the plant reactor is capable of maintaining the desired temperature.

6. Use multiple temperature sensors, in different locations in the reactor for rapid exothermic reactions.

This is particularly important if the reaction mixture contains solids, is very viscous, or if the reactor has coils or other internal elements which might inhibit good mixing.

7. Avoid feeding a material to a reactor at a higher temperature than the boiling point of the reactor contents.

This can cause rapid boiling of the reactor contents and vapor generation.

C.3 Resources and Publications

There are many valuable books and other resources to help in understanding and managing reactive chemistry hazards. Some particularly useful resources include:

- American Institute of Chemical Engineers, Center for Chemical Process Safety, *Safety* Alert: Reactive Material Hazards, New York, 2001.
- Bretherick's Handbook of Reactive Chemical Hazards, 7th Ed., Butterworth-Heineman, 2007.
- *Chemical Reactivity Worksheet*, U. S. National Oceanic and Atmospheric Administration, http://response.restoration.noaa.gov/chemaids/react.html
- American Institute of Chemical Engineers, Center for Chemical Process Safety, Guidelines for Safe Storage and Handling of Reactive Materials, 1995.
- American Institute of Chemical Engineers, Center for Chemical Process Safety, Guidelines for Chemical Reactivity Evaluation and Application to Process Design, 1995.
- United Kingdom Health and Safety Executive, *Designing and Operating Safe* Chemical Reaction Processes, 2000.
- Barton, J., and R. Rogers, Chemical Reaction Hazards: A Guide to Safety, Gulf Publishing Company, 1997.

- Johnson, R. W., S. W. Rudy, and S. D. Unwin. *Essential Practices for Managing Chemical Reactivity Hazards*. New York: American Institute of Chemical Engineers, Center for Chemical Process Safety, 2003.

APPENDIX D – LIST OF SACHE COURSES

Table D.1 List of SACHE Courses

Course	Year
Safety Valves: Practical Design Practices for Relief Valve Sizing	2003
Mini-Case Histories Monsanto polystyrene batch runaway	2003
Mini-Case Histories Morton	2003
The Bhopal disaster: A Case History (2010) & Bhopal - Mini-Case Histories	2003
Mini-Case Histories Flixboro	2003
Mini-Case Histories Hickson decomposition in batch dist unit	2003
Phillips - Mini-Case Histories	2003
Mini-Case Histories Sonat - manual valve alignment	2003
Mini-Case Histories Tosco - refinery release during maintenance	2003
Hydroxylamine Explosion Case Study	2003
Green Engineering Tutorial	2004
Metal Structured Packing Fires	2004
Consequence Modeling Source Models I: Liquids & Gases	2004
Chemical Reactivity Hazards	2005
Introduction to Biosafety	2005
Emergency Relief System Design for Single and Two-Phase Flow	2005
Runaway Reactions -- Experimental Characterization and Vent Sizing	2005
Simplified Relief System Design Package	2005
University Access to SuperChems and ioXpress	2005
Inherently Safer Design	2006
Dust Explosion Prevention and Control	2006
Design for Overpressure and Underpressure Protection	2006
Properties of Materials	2007
Static Electricity I -- Everything You Wanted to Know about Static Electricity	2007
CCPS Process Safety Beacon Archive	2007
Venting of Low Strength Enclosures	2007
Rupture of a Nitroaniline Reactor	2007
Piper Alpha Lessons Learned	2007
Inherently Safer Design Conflicts and Decisions	2008
Static Electricity as an Ignition Source	2008
Risk Assessment	2008
Seminar on Tank Failures	2008
Seveso Accidental Release Case History	2008
Explosions	2009
Reactive and Explosive Materials	2009

Table D.1 List of SACHE Courses, continued

Course	Year
Seminar on Fire	2009
Process Hazard Analysis: An Introduction	2009
Process Hazard Analysis: Process and Examples	2009
Project Risk Analysis (PRA): Unit Operations Lab Applications	2009
Fire Protection Concepts	2010
Process Safety Course Presentations	2010
Safe Handling Practices: Methacrylic Acid	2010
Understanding Atmospheric Dispersion of Accidental Releases	2010
Dow Fire and Explosion Index (F&EI) and Chemical Exposure Index (CEI) Software	2011
Layer of Protection Analysis - Introduction	2011
Compressible and Two-Phase Flow with Applications Including Pressure Relief System Sizing	2011
Case History: A Batch Polystyrene Reactor Runaway	2011
A Process Safety Management, PSM Overview	2012
Conservation of Life: Application of Process Safety Management	2012
T2 Runaway Reaction and Explosion	2012

APPENDIX E – Reactivity Hazard Evaluation Tools

E.1 Screening Table and Flowchart

Table E.1 can be used as a form to document answers to the screening questions described in *Essential Practices for Managing Chemical Reactivity Hazards* (Ref. E.1).

Table E.1 Example Form to Document Screening of Chemical Reactivity Hazards

FACILITY:	COMPLETION DATE:
COMPLETED BY:	APPROVED BY:

Do the answers to the following questions indicate chemical reactivity hazard(s) are present? [1]

AT THIS FACILITY:	YES, NO or NA	BASIS FOR ANSWER; COMMENTS
Question 1. Is intentional chemistry performed?		
2. Is there any mixing or combining of different substances?		
3. Does any other physical processing of substances occur?		
4. Are there any hazardous substances stored or handled?		
5. Is combustion with air the only chemistry intended?		
6. Is any heat generated during the mixing or physical processing of substances?		
7. Is any substance identified as spontaneously combustible?		
8. Is any substance identified as peroxide forming?		
9. Is any substance identified as water reactive?		
10. Is any substance identified as an oxidizer?		
11. Is any substance identified as self-reactive?		
12. Can incompatible materials coming into contact cause undesired consequences, based on the following analysis?		

Table E.1 Example Form to Document Screening of Chemical Reactivity Hazards, continued.

SCENARIO	CONDITIONS NORMAL?²	R, NR or ?³	INFORMATION SOURCES COMMENTS
1			
2			
3			

[1] Use Figure 3.1 with answers to Questions 1-12 to determine if answer is YES or NO

[2] Does the contact/mixing occur at ambient temperature, atmospheric pressure, 21% oxygen atmosphere, and unconfined? (IF NOT, DO NOT ASSUME THAT PUBLISHED DATA FOR AMBIENT CONDITIONS APPLY)

[3] **R** = Reactive (incompatible) under the stated scenario and conditions

NR = Non-reactive (compatible) under the stated scenario and conditions

? = Unknown; assume incompatible until further information is obtained

Figure E.1 is a flowchart that shows how the questions in Table E.1 are connected to determine whether a chemical reactivity hazard can be expected in your facility. The note about Chapter 3 and question numbers refers to the chapters in *Essential Practices for Managing Chemical Reactivity Hazards.*

APPENDIX E

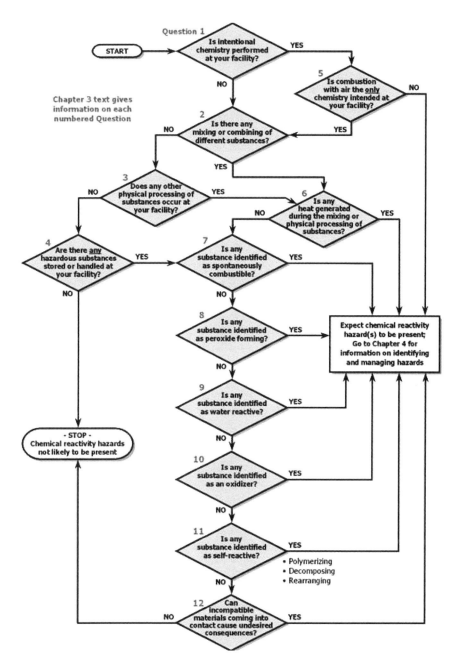

Figure E.1 Summary flowchart, preliminary screening for chemical reactivity hazard analysis.

E.2 Reference

E.1 Essential Practices for Managing Chemical Reactivity Hazards, American Institute of Chemical Engineers, Center of Chemical Process Safety, New York, NY, 2003.

INDEX

A

Accident Prevention Pillar
    Commit to process safety, 12
    Learn from experience, 42
    Manage Risk, 25
    Understand Hazards and Risks, 20
Adsorption hazards, 149
Ammonium Nitrate Explosion
    Port Neal, 25, 77
    Texas City, 38
ARCO Channelview Explosion, 64
Asset Integrity and Reliability, 28, 63, 90, 128
Auditing, 46

B

Bhopal, 6, 110
BLEVE, 32, 175
    Lyon, France Incident, 32
BP Refinery Explosion, 2, 58
Buncefield Explosion, 168

C

Carbon Bed Incident, 151
Centrifugal pumps, 136
    magnetic drive, 137
    seals, 136
Challenger Space Shuttle, 12
Chemical Safety Board, 53
Chevron Richmond Refinery fire, 28
Codes and Standards, 16, 99
Columbia Space Shuttle, 42, 66
Compliance with standards, 109
Compliance with Standards, 15, 87, 124
    Chemical Engineers, 125
    Mechanical Engineers, 125
Concept Sciences explosion, 20, 70
Conduct of Operations, 37, 94, 105, 218
    Engineering Discipline, 230
    Management Discipline, 232
    New Engineers, 237
    Operational Discipline, 37, 218
Continuous improvement, 14
Contractor Management, 30, 87
CSB videos, 131

D

Distillation, 146
    Distillation Column Incident, 150
    Packing material fires, 149
Dust Explosions, 15, 154

E

Emergency Management, 38, 84, 101
Emergency Response, 95
Equilon Anacortes Coking Accident, 191
Extractor Hazards, 149
Exxon Valdez, 102

F

Failure to Learn, 118
Fired Equipment, 163
    Design Considerations, 166
    Example Incidents, 163
Flixborough Explosion, 33, 95
Formosa Plastics VCM Explosion, 91

H

Hazard Identification and Risk Analysis, 22, 73, 76, 80, 111, 126

Heat Exchange Equipment, 141
    Design Considerations, 144
    Example Incidents, 143
HF Release
    Texas City, 188
HIRA, 22, 24
Hydroxylamine explosion, 20, 70

I

Improvement, 14
Incident Investigation, 42, 118, 130
Inherent Safety, 111
    Strategy, 111, 133
Inherently Safer Design, 133, 199

L

Longford gas plant explosion, 46, 73
LPG, 32, 175

M

Macondo Well Blowout, 113
Management of Change, 33, 63, 76, 94, 98, 105, 112, 126
Management of Organizational Change, 34, 127
Management Review and Continuous Improvement, 48
Mars Climate Orbiter, 45
Measurement and Metrics, 45
Mechanial Separation
    Design Considerations, 155
Mechanical Separation, 152
    Batch Centrifuge Incident, 153
    Dust Collector Explosion, 154
Motiva Enterprises explosion, 26

N

Near Miss, 43
New Skills, 215
    Non-Technical, 215
    Technical, 216
Normalization of deviance, 14

O

Operating Procedures, 25, 80
Operational Readiness, 35
Organizational Change Management, 34
OSHA PSM, 8
    elements, 10

P

Partridge Raleigh Oilfield Explosion, 85
PEMEX LPG explosion, 106
Petroleum Processin
    Crude Handling & Separation, 182
Petroleum Processing, 179
    Alkylation, 188
    Catalytic Cracking, 185
    Coking, 190
    Hydrotreating, 184
    Light Hydrocarbon Handling, 183
    Process Safety Hazards, 180, 186, 187
    Process Safety Incidents and Hazards, 183, 184
    Process Safety Incidents and Hazards, 182
    Process Safety Incidents and Hazards, 188
    Reforming, 187
Piper Alpha explosion, 30, 80
Pre- Startup Safety Reviews, 35
Process Knowledge Management
    I&E and Electrical Engineers, 124
    Mechanical Engineers, 124
    Safety Engineers, 124
Process Knowledge Management, 20, 62, 73, 121
    Chemical Engineers, 121
Process Safety, 2
    Design strategies, 133

INDEX

Metric Pyramid, 44
Metrics, 43
New engineer roles, 122, 123
Process Safety Competency, 17, 77
Process Safety Culture, 12, 61, 66, 69, 111, 118
Process Safety Event, 43
Process Safety Incidents, 43
   Swiss Cheese Model, 56
Process Safety Management
   Definition, 8
   Overview Course, 200
Pumps, compressors, fans, 134
   Centrifugal pumps, 136
   Design considerations, 136
   Example Incidents, 135
   Positive displacement pumps, 138

R

RAGEGEP, 16
RBPS Elements
   Asset Integrity and Reliability, 28
   Auditing, 46
   Compliance with Standards, 15
   Conduct of Operations, 37
   Contractor Management, 30
   Emergency Management, 38
   Hazard Identification and Risk Analysis, 22
   Incident Investigation, 42
   Management of Change, 33
   Management Review and Continuous Improvement, 48
   Measurement and Metrics, 45
   Operating Procedures, 25
   Operational Readiness, 35
   Process Knowledge Management, 20
   Process Safety Competency, 17
   Process Safety Culture, 12
   Safe Work Practices, 26
   Stakeholder Outreach, 19
   Training and Performance Assurance, 32
   Workforce Involvement, 18
Reactive Hazards, 158
   Seveso, 159
   T2 Laboratories incident, 160
Reactors, 158
   Design Considerations, 162
Recognized and Generally Accepted Good Engineering Practices, 15
Refinery Hydrocracker Explosion, 37
Risk, 23
   Definition, 9
Risk Based Process Safety
   RBPS Elements, 10

S

SACHE, 199
   Case Histories, 206
   Chemical Reactivity Hazards, 201
   Emergency Relief Systems, 205
   Fires and Explosions, 202
   Hazard Identification and Risk Analysis, 203
   Other Hazards, 203
   Other Modules, 209
   Process Hazards, 201
   Process Safety Overview, 200
Safe Work Practices, 26, 84, 86, 129
   Electrical Engineers, 129
   Instrument Engineers, 129
   Mechanical Engineers, 129
Safety Culture, 217
Safety Training Class Matrix, 212
Sandoz Warehouse Fire, 99
Sense of vulnerability, 13, 118, 217
Seveso, 159
Seveso Directive, 5
Stakeholder Outreach, 19
Storage, 167
   Buncefield Explosion, 168

Design Considerations, 171
General failure modes, 176
Pressurized storage tanks, 175
Underground Storage Tanks, 171
Vacuum Collapse, 171
Wrong Chemical Unloaded, 168

T

Texaco, Milford Haven explosion, 88

Training and performance assurance, 62
Training and Performance Assurance, 32
Transient Operating States, 192
   Design Considerations, 194
   Process Safety Incidents, 192

W

Workforce Involvement, 18